Titles in This Series

Volume

1 **Markov random fields and their applications,** Ross Kindermann and J. Laurie Snell

2 **Proceedings of the conference on integration, topology, and geometry in linear spaces,** William H. Graves, Editor

3 **The closed graph and P-closed graph properties in general topology,** T. R. Hamlett and L. L. Herrington

4 **Problems of elastic stability and vibrations,** Vadim Komkov, Editor

5 **Rational constructions of modules for simple Lie algebras,** George B. Seligman

6 **Umbral calculus and Hopf algebras,** Robert Morris, Editor

7 **Complex contour integral representation of cardinal spline functions,** Walter Schempp

8 **Ordered fields and real algebraic geometry,** D. W. Dubois and T. Recio, Editors

9 **Papers in algebra, analysis and statistics,** R. Lidl, Editor

10 **Operator algebras and K-theory,** Ronald G. Douglas and Claude Schochet, Editors

11 **Plane ellipticity and related problems,** Robert P. Gilbert, Editor

12 **Symposium on algebraic topology in honor of José Adem,** Samuel Gitler, Editor

13 **Algebraists' homage: Papers in ring theory and related topics,** S. A. Amitsur, D. J. Saltman, and G. B. Seligman, Editors

14 **Lectures on Nielsen fixed point theory,** Boju Jiang

15 **Advanced analytic number theory. Part I: Ramification theoretic methods,** Carlos J. Moreno

16 **Complex representations of GL(2, K) for finite fields K,** Ilya Piatetski-Shapiro

17 **Nonlinear partial differential equations,** Joel A. Smoller, Editor

18 **Fixed points and nonexpansive mappings,** Robert C. Sine, Editor

19 **Proceedings of the Northwestern homotopy theory conference,** Haynes R. Miller and Stewart B. Priddy, Editors

20 **Low dimensional topology,** Samuel J. Lomonaco, Jr., Editor

21 **Topological methods in nonlinear functional analysis,** S. P. Singh, S. Thomeier, and B. Watson, Editors

22 **Factorizations of $b^n \pm 1$, $b = 2, 3, 5, 6, 7, 10, 11, 12$ up to high powers,** John Brillhart, D. H. Lehmer, J. L. Selfridge, Bryant Tuckerman, and S. S. Wagstaff, Jr.

23 **Chapter 9 of Ramanujan's second notebook—Infinite series identities, transformations, and evaluations,** Bruce C. Berndt and Padmini T. Joshi

24 **Central extensions, Galois groups, and ideal class groups of number fields,** A. Fröhlich

25 **Value distribution theory and its applications,** Chung-Chun Yang, Editor

26 **Conference in modern analysis and probability,** Richard Beals, Anatole Beck, Alexandra Bellow, and Arshag Hajian, Editors

27 **Microlocal analysis,** M. Salah Baouendi, Richard Beals, and Linda Preiss Rothschild, Editors

28 **Fluids and plasmas: geometry and dynamics,** Jerrold E. Marsden, Editor

29 **Automated theorem proving,** W. W. Bledsoe and Donald Loveland, Editors

30 **Mathematical applications of category theory,** J. W. Gray, Editor

31 **Axiomatic set theory,** James E. Baumgartner, Donald A. Martin, and Saharon Shelah, Editors

32 **Proceedings of the conference on Banach algebras and several complex variables,** F. Greenleaf and D. Gulick, Editors

Titles in This Series

Volume

33 **Contributions to group theory**, Kenneth I. Appel, John G. Ratcliffe, and Paul E. Schupp, Editors

34 **Combinatorics and algebra**, Curtis Greene, Editor

35 **Four-manifold theory**, Cameron Gordon and Robion Kirby, Editors

36 **Group actions on manifolds**, Reinhard Schultz, Editor

37 **Conference on algebraic topology in honor of Peter Hilton**, Renzo Piccinini and Denis Sjerve, Editors

38 **Topics in complex analysis**, Dorothy Browne Shaffer, Editor

39 **Errett Bishop: Reflections on him and his research**, Murray Rosenblatt, Editor

40 **Integral bases for affine Lie algebras and their universal enveloping algebras**, David Mitzman

41 **Particle systems, random media and large deviations**, Richard Durrett, Editor

42 **Classical real analysis**, Daniel Waterman, Editor

43 **Group actions on rings**, Susan Montgomery, Editor

44 **Combinatorial methods in topology and algebraic geometry**, John R. Harper and Richard Mandelbaum, Editors

45 **Finite groups–coming of age**, John McKay, Editor

46 **Structure of the standard modules for the affine Lie algebra $A_1^{(1)}$**, James Lepowsky and Mirko Primc

47 **Linear algebra and its role in systems theory**, Richard A. Brualdi, David H. Carlson, Biswa Nath Datta, Charles R. Johnson, and Robert J. Plemmons, Editors

48 **Analytic functions of one complex variable**, Chung-chun Yang and Chi-tai Chuang, Editors

49 **Complex differential geometry and nonlinear differential equations**, Yum-Tong Siu, Editor

50 **Random matrices and their applications**, Joel E. Cohen, Harry Kesten, and Charles M. Newman, Editors

51 **Nonlinear problems in geometry**, Dennis M. DeTurck, Editor

52 **Geometry of normed linear spaces**, R. G. Bartle, N. T. Peck, A. L. Peressini, and J. J. Uhl, Editors

53 **The Selberg trace formula and related topics**, Dennis A. Hejhal, Peter Sarnak, and Audrey Anne Terras, Editors

54 **Differential analysis and infinite dimensional spaces**, Kondagunta Sundaresan and Srinivasa Swaminathan, Editors

55 **Applications of algebraic K-theory to algebraic geometry and number theory**, Spencer J. Bloch, R. Keith Dennis, Eric M. Friedlander, and Michael R. Stein, Editors

56 **Multiparameter bifurcation theory**, Martin Golubitsky and John Guckenheimer, Editors

57 **Combinatorics and ordered sets**, Ivan Rival, Editor

58 **The Lefschetz centennial conference. Part I: Proceedings on algebraic geometry**, D. Sundararaman, Editor

59 **Function estimates**, J. S. Marron, Editor

Function Estimates

CONTEMPORARY MATHEMATICS

Volume 59

Function Estimates

Proceedings of a conference held July 28–August 3, 1985

J. S. Marron, Editor

AMERICAN MATHEMATICAL SOCIETY

Providence · Rhode Island

The Conference on Function Estimates was held at Humboldt State University, Arcata, California, July 28 to August 3, 1985. It was sponsored by the American Mathematical Society, the Institute of Mathematical Statistics, and the Society for Industrial and Applied Mathematics. It was supported by the National Science Foundation, Grant DMS-8415201

1980 *Mathematics Subject Classifications* (1985 *Revision*). 62G05, 62G99.

Library of Congress Cataloging-in-Publication Data

Conference on Function Estimates (1985: Humboldt State University)
Function Estimates.
(Contemporary mathematics / American Mathematical Society, ISSN 0271-4132; v. 59)
The Conference was sponsored by the American Mathematical Society, the Institute of Mathematical Statistics, and the Society for Industrial and Applied Mathematics.
Bibliography: p.
1. Estimation theory–Congress. I. Marron, James Stephen, 1954– . II. American Mathematical Society. III. Institute of Mathematical Statistics. IV. Society for Industrial and Applied Mathematics. V. Title. VI. Series: Contemporary mathematics (American Mathematical Society); v. 59.
QA276.8.C65 1985 519.5′44 86-14203
ISBN 0-8218-5062-8 (alk. paper)

Printed in the United States of America.
This volume was printed directly from author-prepared copy.
The paper used in this book is acid-free and falls within the guidelines established to ensure permanence and durability.

CONTENTS

PREFACE

This is the volume of the proceedings of the conference "Function Estimates", organized by Murray Rosenblatt, which took place July 28 to August 3, 1985, at Humboldt State University, Arcata, California. The conference was funded by the National Science Foundation and sponsored by the American Mathematical Society, the Institute of Mathematical Statistics, and the Society for Industrial and Applied Mathematics.

There were 36 participants, of which 16 people gave presentations. The papers in this volume appear in the same order as the talks were presented. The subject matter is very broad, including work in geophysics, numerical analysis and nonparametric statistics. The underlying theme is function estimation, especially in the context of various types of spline estimation and deconvolution problems.

An especially important, unusual and worthwhile feature of this conference was the many opportunities for informal interaction among the participants. Additional thanks are due to C. J. Stone, the third member of the organizing committee, to M. R. Leadbetter for initially suggesting this conference, and to Carole Kohanski for the work she put in with the actual arrangements.

J. S. Marron

Contemporary Mathematics
Volume 59, 1986

LOGSPLINE DENSITY ESTIMATION

Charles J. Stone and Cha-Yong Koo[1]

ABSTRACT. Loglinear models are used to obtain estimates and confidence intervals for an unknown univariate density and for the corresponding quantile function. The emphasis is on cubic spline models for the logarithm of the density. Modifications are introduced to provide for reasonable treatment of tail behavior, especially when the density function is known to be concentrated on the positive half line. The methodology is illustrated with pseudo-random samples from six generalized gamma densities.

1. INTRODUCTION. It is common in data analysis to have a data set consisting of n real numbers $y_1, \cdots, y_n$, with a goal being to determine various statistical quantities such as means, variances, and quantiles. To some extent, a direct empirical approach is possible. We can use the sample mean and sample variance. To obtain the sample quantiles $y_{(q)}$, $0 < q < 1$, we first arrange the data in nondecreasing order $y_{1:n} \leq \cdots \leq y_{n:n}$ and set $y_{(i/(n+1))} = y_{i:n}$ for $1 \leq i \leq n$; we then extend this definition by linear interpolation together with $y_{(q)} = y_{1:n}$ for $q < 1/(n+1)$ and $y_{(q)} = y_{n:n}$ for $q > n/(n+1)$.

In order to create a viable mathematical framework for treating these quantities, we typically make the assumption that $y_1, \cdots, y_n$ are the observed values of a random sample of size n from some fixed underlying distribution function F_0. Although this simple assumption is patently unrealistic in many applications, where there may be obvious correlations between the individual observations or fluctuations in the underlying distribution function, it is clearly a proper starting point for mathematical analysis. It is perhaps surprising that in a century or more of work in statistics, much of it devoted to this setup, there are still fundamental questions that have yet to be treated adequately.

For the purpose at hand, we add the assumption that F_0 is absolutely continuous with density f_0. Then, in principle, knowledge of f_0 allows us

---

1980 *Mathematics Subject Classification* (1985 *Revision*). 62F11, 62G05.

[1]This research was supported in part by NSF Grant MCS 83-01257.

to determine all other aspects of the underlying distribution. In particular, $F_0(y) = \int_{-\infty}^{y} f_0(z)dz$; $y_q = F_0^{-1}(q)$; the mean, $\mu = \int_{-\infty}^{\infty} yf_0(y)dy$; and the variance, $\sigma^2 = \int_{-\infty}^{\infty}(y-\mu)^2 f_0(y)dy$.

Of course, we rarely if ever "know" f_0. But we can construct estimates of it based on the sample data. The classical approach is to start with a parametric model $f(y; \boldsymbol{\theta})$, where $\boldsymbol{\theta}$ is a K-dimensional vector of parameters $\theta_1, \cdots, \theta_K$. Such a model is often called a family or system. Well known examples are normal, lognormal, gamma, Weibull, logistic and Pareto families and the Pearson system. Occasionally, there is some theoretical justification for the choice of the family used. But often the theoretical justification is minimal--we may choose a lognormal, gamma, Weibull or Pareto distribution simply because the data values are necessarily positive and we have a rough idea of the rate of decay of $f_0(y)$ as $y \to \infty$. In a particular field of application the choice may be dictated by tradition, partially justified by previous empirical studies, by mathematical tractability, or by interpretability (for example, the Pareto family could be employed to model the distribution of household income, with the shape parameter being used as a measure of economic inequality). Sometimes the distinction is made between Pearson and similar systems, on the one hand, as "graduating" curves and normal, etc., families, on the other hand, as genuine theoretical models. But there generally is no mathematical basis for such a distinction.

Given a parametric model, the observed data can be used to estimate the vector $\boldsymbol{\theta}$ of its parameters. Here maximum likelihood is commonly employed: choose the value $\hat{\boldsymbol{\theta}}$ of $\boldsymbol{\theta}$ that maximizes the log-likelihood

$$\ell(\boldsymbol{\theta}) = \Sigma_1^n \log f(y_i; \boldsymbol{\theta}).$$

If the model is believed to be exactly valid and has a unique "true" value $\boldsymbol{\theta}_0$, then $\hat{\boldsymbol{\theta}}$ can be viewed as an estimate of $\boldsymbol{\theta}_0$. Under reasonable additional assumptions, for large n, the accuracy of $\hat{\boldsymbol{\theta}}$ can be determined from the formula

$$\mathcal{L}(\hat{\boldsymbol{\theta}} - \boldsymbol{\theta}_0) \doteq \mathcal{N}(0, \mathcal{I}^{-1}(\boldsymbol{\theta}_0)) \doteq \mathcal{N}(0, \mathcal{I}^{-1}(\hat{\boldsymbol{\theta}})). \tag{1}$$

Here $\mathcal{I}(\boldsymbol{\theta})$ is the information matrix, whose (j, k)th element is

$$-n\int \frac{\partial^2 \log f(y; \boldsymbol{\theta})}{\partial\theta_j \partial\theta_k} f(y; \boldsymbol{\theta})dy;$$

and $\mathcal{N}(0, \mathcal{I}^{-1}(\boldsymbol{\theta}))$ is the multivariate normal distribution with mean 0 and covariance matrix $\mathcal{I}^{-1}(\boldsymbol{\theta})$. A quantity defined in terms of $f(\cdot; \boldsymbol{\theta})$ can be written as $g(\boldsymbol{\theta})$. It typically follows from (1) that

(2) $$\mathcal{L}(g(\hat{\boldsymbol{\theta}}) - g(\boldsymbol{\theta}_0)) \doteq \mathcal{N}(0, \nabla g(\hat{\boldsymbol{\theta}}) \cdot \mathcal{I}^{-1}(\hat{\boldsymbol{\theta}}) \nabla g(\hat{\boldsymbol{\theta}})),$$

where ∇ denotes the gradient and "$\cdot$" denotes the usual inner product on K-dimensional Euclidean space. Formula (2) can be used to obtain approximate confidence intervals for $g(\boldsymbol{\theta}_0)$. For example, $g(\hat{\boldsymbol{\theta}}) \pm 1.282\ \mathrm{SE}(g(\hat{\boldsymbol{\theta}}))$ are the endpoints of an approximate 80% confidence interval for $g(\boldsymbol{\theta}_0)$; here the indicated standard error is the positive square root of the expression for the variance in the right side of (2).

Inevitably, however, the model cannot realistically be believed to be exactly valid, so the concept of a "true" parameter $\boldsymbol{\theta}_0$ is questionable. Instead, one typically can define $\boldsymbol{\theta}^*$ as the value of $\boldsymbol{\theta}$ that maximizes

$$\int \log f(y; \boldsymbol{\theta}) f_0(y) dy;$$

then $\hat{\boldsymbol{\theta}}$ and $f(\cdot; \hat{\boldsymbol{\theta}})$ can be viewed as estimates of $\boldsymbol{\theta}^*$ and $f^* = f(\cdot; \boldsymbol{\theta}^*)$ respectively. Formula (2), with $\boldsymbol{\theta}_0$ replaced by $\boldsymbol{\theta}^*$, is typically not valid. But even so, it gives reasonable measures of accuracy if $f^* \doteq f_0$. The difference $f^* - f_0$ is referred to as *model bias*. An obvious way to reduce this bias is to enlarge the model by adding additional parameters.

A seemingly different approach is to eschew a parametric model and employ a nonparametric estimate of f_0. The best known such estimate is the sample histogram. This is defined in terms of the empirical distribution $\hat{P}$ of the original data, given by

$$\hat{P}(A) = n^{-1} \#\{i: 1 \le i \le n \text{ and } y_i \in A\}.$$

Choose a positive integer $N \ge 4$ and define numbers $t_1, \cdots, t_N$ such that $t_1 = y_{1:n} = \min(y_1, \cdots, y_n)$; $t_N = y_{n:n} = \max(y_1, \cdots, y_n)$ and $t_1, \cdots, t_N$ are equally spaced. Set $K = N-1$, $I_K = [t_K, t_{K+1}]$ and $I_k = [t_k, t_{k+1})$ for $1 \le k < K$; and let $|I_k| = t_{k+1} - t_k$ denote the length of the interval I_k. The corresponding histogram density estimate $\hat{f}$ is given by $\hat{f} = \hat{P}(I_k)/|I_k|$ on I_k for $1 \le k \le K$, and $\hat{f} = 0$ on the complement of $[t_1, t_N]$.

If we ignore the dependence of the "knots" $t_1, \cdots, t_N$ on the data, the second approach is seen not to differ in a fundamental way from the first approach. Indeed, let $B_1, \cdots, B_K$ be defined by $B_k = 1$ on I_k and $B_k = 0$ elsewhere. Given the vector $\boldsymbol{\theta}$, define $C(\boldsymbol{\theta})$ by

$$\int_{t_1}^{t_N} \exp(C(\boldsymbol{\theta}) + \Sigma_1^K \theta_k B_k) dy = 1$$

and set

(3) $$f(y; \boldsymbol{\theta}) = \exp(C(\boldsymbol{\theta}) + \Sigma_1^K \theta_k B_k(y))$$

for $y \in [t_1, t_N]$ and $f(y; \theta) = 0$ for $y \notin [t_1, t_N]$. It is easily seen that the maximum likelihood estimate $\hat{\theta}$ of θ exists and is unique and that $\hat{f} = f(\cdot; \hat{\theta})$ coincides with the histogram density estimate previously defined.

A general model of the form (3) is called a *loglinear* model. Such models will be studied in Section 2. Previously Leonard [4] and Silverman [5] have considered other nonparametric models for the logarithm of an unknown density. An attractive feature of any such model is that the density estimate is automatically nonnegative.

The histogram density estimate has two limitations; it is discontinuous at the knots $t_1, \cdots, t_N$; and it does not provide a meaningful way of extrapolating beyond the range of the data. Note that the piecewise constant function $\Sigma_1^K \theta_k B_k$ in (3) is a first order spline. It is natural to replace it by a higher order spline. In Section 3 we will study loglinear models in which $\Sigma_1^K \theta_k B_k$ is a cubic spline. Modifications will be introduced to extrapolate beyond the range of the data and to handle densities which are known to be concentrated on the positive half-line. The resulting methodology will be illustrated in Section 4 with simulated data.

2. LOGLINEAR MODELS. Let $\mathcal{Y}$ be a subset of d-dimensional Euclidean space $\mathbb{R}^d$ having positive (not necessarily finite) Lebesgue measure and let $B_1, \cdots, B_K$ be fixed real valued functions on $\mathcal{Y}$, where $1 \le K < \infty$. Let $\theta_1, \cdots, \theta_K$ denote the coordinates of $\theta \in \mathbb{R}^K$. Also set

$$A(y; \theta) = \Sigma_1^K \theta_k B_k(y).$$

It is assumed that, unless the elements of θ are all zero, $A(y; \theta)$ is not (almost everywhere) equal to a constant on $\mathcal{Y}$. Consider a loglinear model

$$\log f(y; \theta) = C(\theta) + A(y; \theta), \qquad y \in \mathcal{Y},$$

for an unknown density f_0 on $\mathcal{Y}$. Correspondingly

$$f(y; \theta) = \exp(C(\theta) + A(y; \theta)).$$

The constraint that the density integrate to one leads to the formula

$$C(\theta) = -\log(\int_{\mathcal{Y}} \exp(A(y; \theta))dy).$$

Set $\Theta = \{\theta \in R^K: C(\theta) > -\infty\}$. Then Θ is a convex subset of R^K. It is assumed that Θ has a nonempty interior Θ_0. Let $\mathbf{H}(\theta)$, $\theta \in \Theta_0$, denote the Hessian of $C(\theta)$, the $K \times K$ matrix whose (j, k)th element is

$$\frac{\partial^2 C(\theta)}{\partial\theta_j \partial\theta_k} = -\int_{\mathcal{Y}} B_j(y)B_k(y)f(y; \theta)dy + \int_{\mathcal{Y}} B_j(y)f(y; \theta)dy \int_{\mathcal{Y}} B_k(y)f(y; \theta)dy.$$

It is easily seen that if $\boldsymbol{\tau} \in \mathbb{R}^K$, then

$$\boldsymbol{\tau} \cdot \mathbf{H}(\boldsymbol{\theta})\boldsymbol{\tau} = -\int_{\mathcal{Y}}(A(y;\ \boldsymbol{\tau}) - a)^2 f(y;\ \boldsymbol{\theta})dy,$$

where

$$a(\boldsymbol{\tau};\ \boldsymbol{\theta}) = \int_{\mathcal{Y}} A(y;\ \boldsymbol{\tau})f(y;\ \boldsymbol{\theta})dy.$$

Thus $-\mathbf{H}(\boldsymbol{\theta})$ is negative definite and hence $C(\cdot)$ is strictly concave on Θ_0. It is assumed that there is a unique $\boldsymbol{\theta}^* \in \Theta$ that maximizes

$$\int_{\mathcal{Y}} \log f(y;\ \boldsymbol{\theta})f_0(y)dy, \qquad \boldsymbol{\theta} \in \Theta,$$

and that $\boldsymbol{\theta}^* \in \Theta_0$. Set $f^* = f(\cdot;\ \boldsymbol{\theta}^*)$.

Let $Y_1, \cdots, Y_n$ be a random sample of size n from f_0 and let $\ell(\boldsymbol{\theta})$, $\boldsymbol{\theta} \in \Theta$, be the log-likelihood function based on the loglinear model; so that

$$\ell(\boldsymbol{\theta}) = \Sigma_1^n \log f(Y_i;\ \boldsymbol{\theta}) = \Sigma_1^n(C(\boldsymbol{\theta}) + A(Y_i;\ \boldsymbol{\theta})).$$

Let $\hat{\boldsymbol{\theta}}$ denote the maximum of $\ell(\cdot)$ over Θ_0. Then $\hat{\boldsymbol{\theta}}$ is unique if it exists. Let $\hat{f} = f(\cdot;\ \hat{\boldsymbol{\theta}})$ denote the corresponding *loglinear density estimator*.

Let $G(y;\ \boldsymbol{\theta}) = \nabla\log f(y;\ \boldsymbol{\theta})$ denote the gradient of $\log f(y;\ \boldsymbol{\theta})$ (with respect to $\boldsymbol{\theta}$). Then $G(y;\ \boldsymbol{\theta})$ is the K-dimensional vector of elements

$$\frac{\partial C}{\partial \theta_k}(\boldsymbol{\theta}) + B_k(y).$$

Let $\mathbf{S}(\boldsymbol{\theta}) = \Sigma_1^n G(Y_i;\ \boldsymbol{\theta}) = \nabla\ell(\boldsymbol{\theta})$ denote the score function; that is the K-dimensional vector of elements

$$\frac{\partial \ell}{\partial \theta_k}(\boldsymbol{\theta}) = \Sigma_1^n \left[\frac{\partial C}{\partial \theta_k}(\boldsymbol{\theta}) + B_k(Y_i)\right] = n\frac{\partial C}{\partial \theta_k}(\boldsymbol{\theta}) + B_{k\cdot},$$

where the sufficient statistics $B_{1\cdot}, \cdots, B_{K\cdot}$ are defined by

$$B_{k\cdot} = \Sigma_1^n B_k(Y_i).$$

The maximum likelihood equation for $\hat{\boldsymbol{\theta}}$ is $\mathbf{S}(\hat{\boldsymbol{\theta}}) = 0$. Let $\boldsymbol{\mathcal{I}}(\boldsymbol{\theta}) = -n\mathbf{H}(\boldsymbol{\theta})$ denote the information matrix corresponding to the random sample. The Newton-Raphson method for converging to $\hat{\boldsymbol{\theta}}$ is to start with an initial guess $\hat{\boldsymbol{\theta}}^{(0)}$ and iteratively determine $\hat{\boldsymbol{\theta}}^{(m)}$ by the formula

$$\hat{\boldsymbol{\theta}}^{(m+1)} = \hat{\boldsymbol{\theta}}^{(m)} + \boldsymbol{\mathcal{I}}^{-1}(\hat{\boldsymbol{\theta}}^{(m)})\mathbf{S}(\hat{\boldsymbol{\theta}}^{(m)}).$$

Observe that these iterations depend on the data only through the sufficient statistics $B_{1\cdot}, \cdots, B_{K\cdot}$. If, at some stage, $\ell(\hat{\boldsymbol{\theta}}^{(m+1)}) \le \ell(\hat{\boldsymbol{\theta}}^{(m)})$, then $\hat{\boldsymbol{\theta}}^{(m+1)}$ should be replaced by $\hat{\boldsymbol{\theta}}^{(m)} + c\boldsymbol{\mathcal{I}}^{-1}(\hat{\boldsymbol{\theta}}^{(m)})\mathbf{S}(\hat{\boldsymbol{\theta}}^{(m)})$ for some constant

$c \in (0, 1)$. (It may be necessary to repeat this step, replacing c by c^2, c^3, etc. until improvement results. Since $\mathcal{I}(\hat{\boldsymbol{\theta}}^{(m)})$ is positive definite, so is its inverse; consequently,

$$\ell(\hat{\boldsymbol{\theta}}^{(m)} + c\mathcal{I}^{-1}(\hat{\boldsymbol{\theta}}^{(m)})\mathbf{S}(\hat{\boldsymbol{\theta}}^{(m)})) > \ell(\hat{\boldsymbol{\theta}}^{(m)})$$

for sufficiently small $c > 0$ unless $\hat{\boldsymbol{\theta}}^{(m)} = \hat{\boldsymbol{\theta}}$.) If $\mathcal{Y}$ is a compact set, it is reasonable to start with $\hat{\boldsymbol{\theta}}^{(0)} = \mathbf{0}$.

According to the usual asymptotics

$$0 = \mathbf{S}(\hat{\boldsymbol{\theta}}) \doteq \mathbf{S}(\boldsymbol{\theta}^*) + n\mathbf{H}(\boldsymbol{\theta}^*)(\hat{\boldsymbol{\theta}} - \boldsymbol{\theta}^*) = \mathbf{S}(\boldsymbol{\theta}^*) - \mathcal{I}(\boldsymbol{\theta}^*)(\hat{\boldsymbol{\theta}} - \boldsymbol{\theta}^*)$$

and hence

$$\hat{\boldsymbol{\theta}} - \boldsymbol{\theta}^* \doteq \mathcal{I}^{-1}(\boldsymbol{\theta}^*)\mathbf{S}(\boldsymbol{\theta}^*).$$

Consequently,

$$\mathcal{L}(\hat{\boldsymbol{\theta}} - \boldsymbol{\theta}^*) \doteq \mathcal{N}(0, \mathcal{I}^{-1}(\boldsymbol{\theta}^*)\mathrm{Cov}(\mathbf{S}(\boldsymbol{\theta}^*))\mathcal{I}^{-1}(\boldsymbol{\theta}^*)),$$

where $\mathrm{Cov}(\mathbf{S}(\boldsymbol{\theta}^*))$ is the covariance matrix of $\mathbf{S}(\boldsymbol{\theta}^*)$. If $f^* \doteq f_0$, then $\mathrm{Cov}(\mathbf{S}(\boldsymbol{\theta}^*)) \doteq \mathcal{I}(\boldsymbol{\theta}^*)$ and hence

$$\mathcal{L}(\hat{\boldsymbol{\theta}} - \boldsymbol{\theta}^*) \doteq \mathcal{N}(0, \mathcal{I}^{-1}(\boldsymbol{\theta}^*)) \doteq \mathcal{N}(0, \mathcal{I}^{-1}(\hat{\boldsymbol{\theta}})). \tag{4}$$

(Otherwise, we could estimate $\mathrm{Cov}(\mathbf{S}(\boldsymbol{\theta}^*))$ by n times the sample covariance matrix of the n vectors $\mathbf{G}(Y_1; \boldsymbol{\theta}), \cdots, \mathbf{G}(Y_n; \boldsymbol{\theta})$.)

Observe that

$$\log \hat{f}(y) = C(\hat{\boldsymbol{\theta}}) + \Sigma_1^K \hat{\theta}_k B_k(y).$$

It follows from (2) and (4) that

$$\mathcal{L}(\log \hat{f}(y) - \log f^*(y)) \doteq \mathcal{N}(0, \mathbf{G}(y; \hat{\boldsymbol{\theta}})\cdot\mathcal{I}^{-1}(\hat{\boldsymbol{\theta}})\mathbf{G}(y; \hat{\boldsymbol{\theta}})). \tag{5}$$

Set

$$\mathrm{SE}(\log \hat{f}(y)) = (\mathbf{G}(y; \hat{\boldsymbol{\theta}})\cdot\mathcal{I}^{-1}(\hat{\boldsymbol{\theta}})\mathbf{G}(y; \hat{\boldsymbol{\theta}}))^{1/2}.$$

Given $0 < q < 1$, let z_q be defined by $\Phi(z_q) = q$, where Φ is the standard normal distribution function. Then by (5)

$$\log \hat{f}(y) \pm z_{(1+\alpha)/2}\ \mathrm{SE}(\log \hat{f}(y))$$

are the endpoints of an approximate α-level confidence interval for $\log f^*(y)$ and hence

$$\hat{f}(y)\exp(\pm z_{(1+\alpha)/2}\ \mathrm{SE}(\log \hat{f}(y)))$$

are the endpoints of an approximate α-level confidence interval for $f^*(y)$.

Suppose now that $\mathcal{Y}$ is a subinterval of the real line having left endpoint $a \geq -\infty$. Let $F(\cdot; \boldsymbol{\theta})$ denote the distribution function corresponding to

$f(\cdot;\ \boldsymbol{\theta})$; so that

$$F(y;\ \boldsymbol{\theta}) = \int_a^y f(x;\ \boldsymbol{\theta})dx = \int_a^y \exp(C(\boldsymbol{\theta}) + \Sigma_1^K \theta_k B_k(x))dx.$$

The gradient, $\nabla F(y;\ \boldsymbol{\theta})$, of $F(y;\ \boldsymbol{\theta})$ is the K-dimensional vector having elements

$$\frac{\partial F}{\partial \theta_k}(y;\ \boldsymbol{\theta}) = \frac{\partial C}{\partial \theta_k}(\boldsymbol{\theta})\ F(y;\ \boldsymbol{\theta}) + \int_a^y B_k(x)f(x;\ \boldsymbol{\theta})dx.$$

Set $F^* = F(\cdot;\ \boldsymbol{\theta}^*)$, $\hat{F} = F(\cdot;\ \hat{\boldsymbol{\theta}})$ and $\nabla\hat{F}(y) = \nabla F(y;\ \hat{\boldsymbol{\theta}})$. Then

$$\mathcal{L}(\hat{F}(y) - F^*(y)) \doteq \mathcal{N}(0,\ \nabla\hat{F}(y)\cdot\mathcal{I}^{-1}(\hat{\boldsymbol{\theta}})\nabla\hat{F}(y)),$$

which can be used to get approximate confidence intervals for $F^*(y)$.

Given $0 < q < 1$, let $y_q(\boldsymbol{\theta})$ denote the qth quantile of $F(\cdot;\ \boldsymbol{\theta})$; so that $F(y_q(\boldsymbol{\theta});\ \boldsymbol{\theta}) = q$. Then

$$\nabla y_q(\boldsymbol{\theta}) = -\frac{1}{f(y_q;\ \boldsymbol{\theta})}\ \nabla F(y_q;\ \boldsymbol{\theta}).$$

Let $y_q^* = y_q(\boldsymbol{\theta}^*)$ and $\hat{y}_q = y_q(\hat{\boldsymbol{\theta}})$ denote the qth quantiles of F^* and $\hat{F}$ respectively. Write $\nabla y_q(\hat{\boldsymbol{\theta}})$ as $\nabla\hat{y}_q$. Then

$$\mathcal{L}(\hat{y}_q - y_q^*) \doteq \mathcal{N}(0,\ \nabla\hat{y}_q\cdot\mathcal{I}^{-1}(\hat{\boldsymbol{\theta}})\nabla\hat{y}_q),$$

which can be used to get approximate confidence intervals for y_q^*.

3. LOGSPLINE MODELS. It is now supposed that $\mathcal{Y}$ is a nondegenerate subinterval of the real line $\mathbb{R}$. Suppose first that $\mathcal{Y}$ is a bounded interval $[a, b]$. Let $N \ge 4$ and let $a < t_1 < \cdots < t_N < b$. Let $\mathcal{S}$ denote the collection of twice continuously differentiable functions s on $[a, b]$ such that s is a polynomial of degree 3 (or less) on each of the intervals $[a, t_1]$, $[t_1, t_2]$, $\cdots$, $[t_N, b]$. Then $\mathcal{S}$ is the $(N+4)$-dimensional vector space of cubic splines corresponding to the knot positions $t_1, \cdots, t_N$. There is a basis $B_1, \cdots, B_{N+4}$ of $\mathcal{S}$ consisting of B-splines (see de Boor [1]). These are nonnegative functions on $[a, b]$ which sum to one on this interval. We can implement the methodology of Section 2 by deleting one of these B-splines, say the last one, and setting $K = N+3$. The resulting procedure is adequate if the true underlying density f_0 is continuous and nonzero on $[a, b]$. (Under this condition the existence of $\boldsymbol{\theta}^*$ is verified in Stone [6]; there the asymptotic properties of the procedure are treated when n and N both tend to ∞.) But the procedure is inadequate if f_0 vanishes at one or both endpoints; and its direct extension to unbounded intervals is also inadequate.

We will now consider modifications designed to deal adequately with these two exceptions; that is, to estimate in a reasonable manner the tails of a density function as well as its central portion. To this end, let the B-splines first be extended to all of $\mathbb{R}$ so as to be cubic polynomials on $(-\infty, t_1]$ and on $[t_N, \infty]$. This can be done in a unique manner. But the space, again denote it by $\mathcal{S}$, of linear combinations of the resulting functions is too flexible in the tails to provide a good basis for loglinear density estimation. (Too much flexibility leads to too high a variance of estimation.)

Let $\mathcal{S}_0$ denote the subspace consisting of all functions in $\mathcal{S}$ that are linear on $(-\infty, t_1]$ and on $[t_N, \infty)$. We will now determine a basis for $\mathcal{S}_0$. To this end, for $s \in \mathcal{S}$, let $s^{(3)}(t_1)$ and $s^{(3)}(t_2)$ denote the left third derivatives of s at t_1 and t_2 respectively ($s^{(3)}(t_1)$ is the third derivative at t_1 of s viewed as a function on $(a, t_1]$). Similarly, let $s^{(3)}(t_{N-1})$ and $s^{(3)}(t_N)$ denote the right third derivatives of s at t_{N-1} and t_N respectively. Note that $t_2 < t_{N-1}$ since it has been assumed that $N \geq 4$. The B-spline B_1 is a strictly positive cubic polynomial on (a, t_1) and vanishes on $[t_1, \infty)$. Since B_1 is twice continuously differentiable, it follows that $B_1^{(2)}(t_1) = 0$, $B_1^{(3)}(t_1) < 0$, $B_1^{(2)}(t_N) = 0$, and $B_1^{(3)}(t_N) = 0$. Further, B_2 is a strictly positive cubic polynomial on $[t_1, t_2)$ and vanishes on $[t_2, \infty)$. It follows that $B_2^{(2)}(t_2) = 0$ and $B_2^{(3)}(t_2) < 0$. Consequently, $B_2^{(2)}(t_1) > 0$, $B_2^{(2)}(t_N) = 0$, and $B_2^{(3)}(t_N) = 0$. Analogous results hold for B_{N+3} and B_{N+4}. Thus there exist unique linear functions c_1, c_2, c_3, and c_4 on $\mathcal{S}$ such that

$$T(s) = s - c_1(s)B_1 - c_2(s)B_2 - c_3(s)B_{N+3} - c_4(s)B_{N+4} \in \mathcal{S}_0$$

for all $s \in \mathcal{S}$. If $s \in \mathcal{S}_0$, then $c_1(s) = \cdots = c_4(s) = 0$ and hence $T(s) = s$. Therefore, $\mathcal{S}_0 = \{T(s)\colon s \in \mathcal{S}\}$ and $\{T(B_k)\colon 1 \leq k \leq N+4\}$ spans $\mathcal{S}_0$. Moreover, $T(B_k) = 0$ for $k = 1, 2, N+3$, and $N+4$; so that $\{T(B_k)\colon 3 \leq k \leq N+2\}$ is a basis of $\mathcal{S}_0$. Now the constant function 1 on $\mathbb{R}$ is in $\mathcal{S}_0$, so that

$$\Sigma_3^{N+2}\, T(B_k) = \Sigma_1^{N+4}\, T(B_k) = T(\Sigma_1^{N+4}\, B_k) = T(1) = 1.$$

Set $K = N-1$. By a change of notation let the functions $T(B_3), \cdots, T(B_{N+1})$ now be denoted by $B_1, \cdots, B_K$ respectively. It is these functions to which we will apply the methodology of Section 2.

In calculating

$$\frac{\partial C(\boldsymbol{\theta})}{\partial \boldsymbol{\theta}_k} \quad \text{and} \quad \frac{\partial^2 C(\boldsymbol{\theta})}{\partial \boldsymbol{\theta}_j \partial \boldsymbol{\theta}_k}$$

it is necessary to evaluate integrals of the forms

$$\int_{-\infty}^{\infty} B_k(y)\exp(C(\boldsymbol{\theta}) + \Sigma_1^K\, \theta_k B_k(y))dy$$

and

$$\int_{-\infty}^{\infty} B_j(y)B_k(y)\exp(C(\boldsymbol{\theta}) + \Sigma_1^K \theta_k B_k(y))dy.$$

Write

$$\int_{-\infty}^{\infty} = \int_{-\infty}^{t_1} + \int_{t_1}^{t_2} + \cdots + \int_{t_{N-1}}^{t_N} + \int_{t_N}^{\infty}.$$

The integrals over $(-\infty, t_1]$ and over $[t_N, \infty)$ can be evaluated analytically since each B_k is linear over these two intervals. The integrals over each of the bounded intervals can be evaluated by Simpson's rule. To obtain the maximum likelihood estimate $\hat{\boldsymbol{\theta}}$ it is appropriate first to restrict y to a bounded interval containing $[t_1, t_N]$ and all the sample data, obtaining the maximum likelihood estimate $\hat{\boldsymbol{\theta}}^{(0)}$; then $\hat{\boldsymbol{\theta}}^{(0)}$ can be used as the starting value in the Newton-Raphson method for obtaining the maximum likelihood estimate $\hat{\boldsymbol{\theta}}$ corresponding to $\mathcal{Y} = \mathbb{R}$. (The bounded interval must be sufficiently large so that $\Sigma_1^K \hat{\theta}_k^{(0)} B_k$ has positive slope at t_1 and negative slope at t_N.)

As just described, logspline density estimation is appropriate if both tails of f_0 decrease at some exponential rate or even if both tails of $\log f_0$ decrease at some moderate algebraic rate. Consider, however, a generalized gamma density

$$\begin{aligned} f_0(y) &= cy^{\alpha-1}\exp(-(y/\beta)^{\gamma}), \quad & y > 0, \\ &= 0, & y \le 0, \end{aligned}$$

where c is simply defined in terms of the positive constants α, β, and γ. Logspline density estimation, as described, can be expected to handle the right tail of f_0 adequately for moderate values of γ but not the left tail for moderate values of α. Let Y be a random variable having density $f_Y = f_0$. It is better to apply logspline density estimation to a transformed random variable $X = T(Y)$ for some strictly increasing function T that maps $(0, \infty)$ onto $(-\infty, \infty)$. The transformation $T(y) = \log(y)$ adequately takes care of the left tail but yields a poor estimate of the right tail. A better choice is given by

$$T(y) = (1-c)\log y + cy$$

for some number $c \in (0, 1)$. It is simple and seems reasonable to use $c = .5$ when $\mathcal{L}(Y)$ has median 1. This leads to the choice of

$$T(y) = .5 \log \frac{y}{\mathrm{Median}(\mathcal{L}(Y))} + .5 \frac{y}{\mathrm{Median}(\mathcal{L}(Y))},$$

which, up to an affine transformation, is of the form

$$T(y) = (1-c)\log y + cy \quad \text{with} \quad c = 1/(1 + \mathrm{Median}(\mathcal{L}(Y))).$$

In practice, we choose $c = 1/(1 + \text{median}(y_1, \cdots, y_n))$.

Observe that

$$f_Y(y) = f_X((1-c)\log y + cy)(\frac{1-c}{y} + c).$$

Thus

$$\hat{f}_Y(y) = \hat{f}_X((1-c)\log y + cy)(\frac{1-c}{y} + c)$$

and approximate confidence intervals for f_X lead similarly to approximate confidence intervals for f_Y. Write the confidence interval for $f_Y(y)$ in the form $[\underline{f}_Y(y), \overline{f}_Y(y)]$. It is easily seen that as $y \to \infty$

$$\hat{f}_Y(y) \sim a_0 y^{b_0}, \qquad \underline{f}(y) \sim a_1 y^{b_1}, \qquad \text{and} \qquad \overline{f}(y) \sim a_2 y^{b_2};$$

here the a's and b's are random constants such that the a's are all positive and $-1 < b_1 < b_0 < b_2$.

Observe that $T(y_q) = x_q$ and hence that $y_q = T^{-1}(x_q)$ and $\hat{y}_q = T^{-1}(\hat{x}_q)$. (The inverse function T^{-1} can be determined by Newton's method). Similarly, approximate confidence intervals for x_q lead to corresponding intervals for y_q. For q close to 1, however, the resulting intervals are nearly symmetric about $\hat{y}_q$. Based on Breiman and Stone [2], we would expect that the upper bound of such an interval would be too low to have the proper coverage probabilities for an $(1+\alpha)/2$-level upper confidence bound and that the lower bound of the interval would similarly be too low. To obtain a more promising confidence interval procedure for y_q, let s denote the standard error of $\hat{x}_q$. Then the standard error of $\log \hat{y}_q$ is

$$t = \frac{s}{\hat{y}_q \, T'(\hat{y}_q)} = \frac{s}{c\hat{y}_q + 1 - c}.$$

This leads to

$$\hat{y}_q \exp\left\{\pm z_{(1+\alpha)/2} \frac{s}{c\hat{y}_q + 1 - c}\right\}$$

as the endpoints of an approximate α-level confidence interval for y_q.

It is necessary, of course, to select N and the knot locations $t_1, \cdots, t_N$. Based on limited simulations with regular densities (generalized gamma and lognormal), we have come up with the choice $N = 5$ (so that $K = 4$) and the following automatic rule for choosing the five knot positions: $t_j = x_{(q)}$ with $q = q_j$ for $j = 1, \cdots, 5$ defined so that $q_1 = 1/(n+1)$, $q_5 = n/(n+1)$, and the numbers $\text{logit}(q_j)$, $1 \le j \le 5$, are equally spaced, where $\text{logit}(q) = \log(q/(1-q))$. The choices of $t_1 = x_{1:n} = \min(x_1, \cdots, x_n)$ and $t_5 = x_{n:n} = \max(x_1, \cdots, x_n)$ were motivated in part by the definition of a "natural" spline. Note also that $t_3 = x_{(.5)} = \text{median}(x_1, \cdots, x_n)$, which equals $x_{(n+1)/2:n}$ or $(x_{n/2:n} + x_{n/2+1:n})/2$ according as n is odd or even.

The choice of a 4-parameter system ($K = 4$) also seems reasonable. Johnson [3] states that "Provided that a suitable form of curve has been chosen, fitting a maximum of *four* parameters gives a reasonably effective approximation." One can think of one parameter for location, one for scale, one for shape of the left tail, and one for shape of the right tail. The well known Pearson system also has four free parameters, corresponding to location, scale, skewness, and kurtosis. There is another similarity between the 4-parameter logspline system and the Pearson system. The former system models $\log f_0$ as a spline, while the latter system models the first derivative of $\log f_0$ as a rational function of the explict form $-c(y+a_0)/(y^2+a_1y+a_2)$.

In order to model accurately a multimodal or otherwise irregular distribution, it may be necessary to use more than four free parameters. This is easily done by choosing $N > 5$. Alternatively, one can keep the number of knots fixed at five, but attempt to choose their positions by maximum likelihood. This is not computationally attractive, however, since the log-likelihood is undoubtedly not a concave function of the knot positions. The conceptual and computational problems increase if we attempt to choose the number of knots as well as their positions in some "optimal" manner. (This is not to say that no such procedure would be practicable.)

4. EXAMPLES. The 4-parameter logspline density estimation procedure described in Section 3 was applied to six generalized gamma distributions. Let Z have a gamma distribution with shape parameter $\alpha = 2$ and scale parameter $\lambda = 1$. Then Z is distributed as the sum of two independent exponential random variables each having mean one; and Z has density f_Z given by $f_Z(z) = ze^{-z}$ for $z > 0$ and $f_Z(z) = 0$ for $z \le 0$. The distribution function of Z is given explicitly by $F_Z(z) = 1 - e^{-z}(1+z)$ for $z > 0$ and $F_Z(z) = 0$ for $z \le 0$; from this the quantiles z_q of Z are easily determined by Newton's method. Set $Y = Z^\beta$ for some $\beta > 0$. Then

$$f_Y(y) = \beta^{-1}y^{2\beta^{-1}-1}\exp(-y^{\beta^{-1}}) \qquad \text{for} \quad y > 0$$

and $f_Y(y) = 0$ for $y \le 0$; also $y_q = z_q^\beta$. The six choices of β that we used are: $1/2$, $4^{.2}/2$, $4^{.4}/2$, $4^{.6}/2$, $4^{.8}/2$, and 2. The corresponding approximate numerical values are: .5, .66, .87, 1.15, 1.52, and 2. For the first five choices of β, $f_Y(0+) = 0$; while for the last choice, $f_Y(0+) = .5$.

We chose $n = 200$ for the sample size and applied the transformation $X = (1-c)\log Y + cY$, where $c = 1/(1 + \text{median}(y_1, \cdots, y_n))$. In applying logspline density estimation to the observed data from $\mathcal{L}(X)$, we chose the five knots in accordance with the automatic rule described in Section 3: $x_{1:200}$; $.7x_{13:200} + .3x_{14:200}$; $.5x_{100:200} + .5x_{101:200}$; $.3x_{187:200} + .7x_{188:200}$; and $x_{200:200}$. We obtained confidence intervals for $f_Y(y)$ and y_q by first obtaining the corresponding intervals for $\log f_X(x)$ and $\log y_q$ respectively and then transforming.

Figure 1 (on two pages) shows the results. In each of the six pairs of plots the left plot refers to density estimation and the right plot to quantile estimation. For the graphs in the density plots, y ranges from $\min(y_1, \cdots, y_n)$ to $\max(y_1, \cdots, y_n)$. In the quantile plots, the first coordinate is $\text{logit}(q)$ but the labels show the values of q corresponding to the various tick marks. The solid lines in the plots show 90% lower confidence bounds and 90% upper confidence bounds and thereby determine 80% confidence intervals, while the dashed lines show the true functions. In the quantile plots, the 10 lowest and 10 highest data values are indicated by asterisks. The estimates themselves are not shown. For in all cases they are only slightly below the average of the corresponding upper and lower confidence bounds. We were pleasantly surprised by the reasonableness of the confidence bounds shown in Figure 1 and in plots based on other simulated data, especially by the appearance of the confidence bounds for y_q with q very near 1. (For example, the 90% upper confidence bound for $y_{.999}$ when $\beta = 2$ lies below the true value apparently because the 10 highest data values are unusually low, not because of model bias.)

We have not yet performed a Monte Carlo study to determine the actual coverage probabilities of these confidence bounds or to investigate alternate choices of the number and position of knots. In doing so, we could reduce the amount of computation substantially by using $\boldsymbol{\theta}^*$ as the starting value in the Newton-Raphson method for determining the maximum likelihood estimate $\hat{\boldsymbol{\theta}}$. It would be interesting to compare the performance of the confidence bounds for y_q with q near 1 obtained via logspline density estimation with those obtained by the quadratic tail procedure in Breiman and Stone [2].

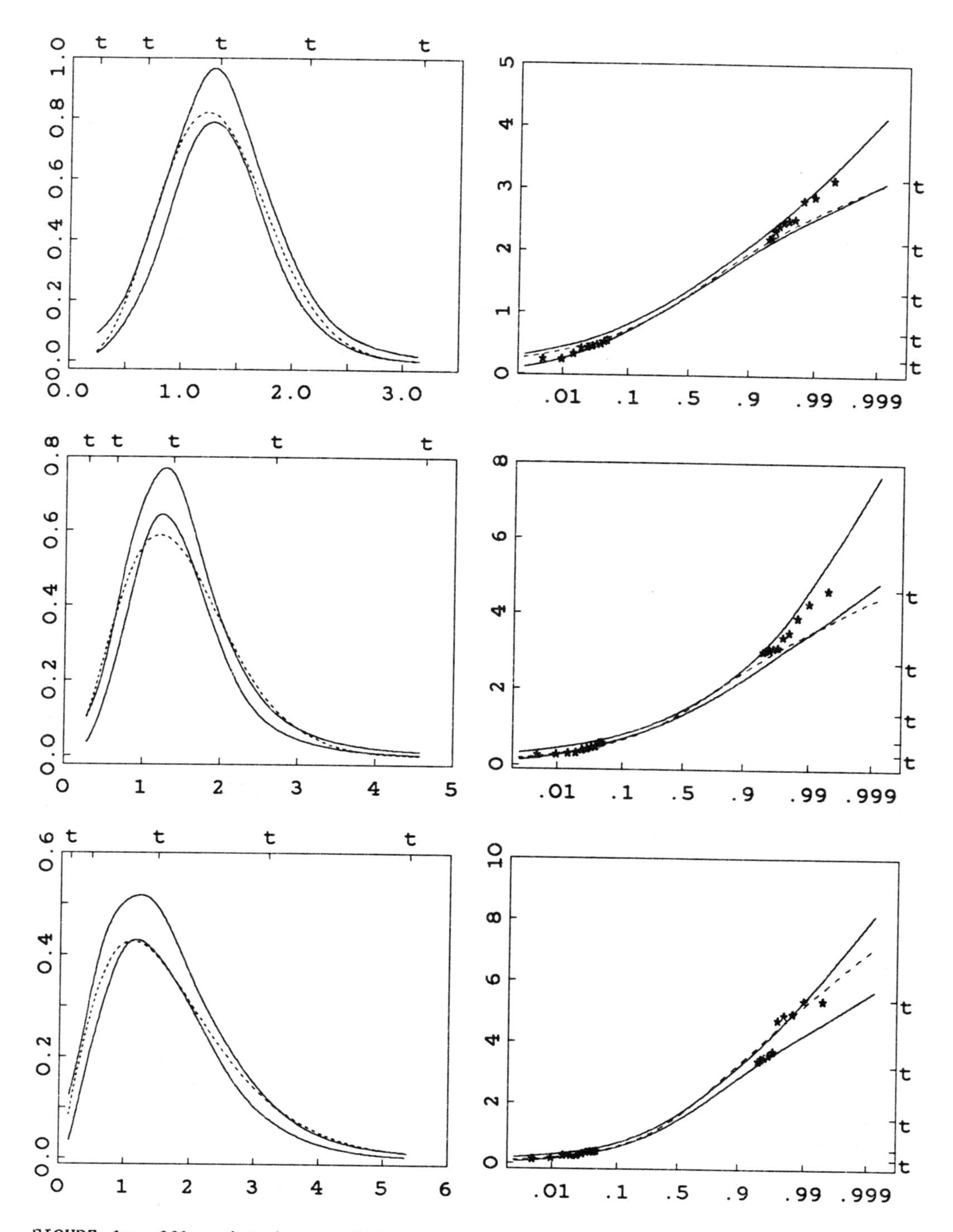

FIGURE 1a. 80% pointwise confidence intervals for density functions, left, and quantiles, right. From top to bottom $\beta \doteq .5, .66, .87$.

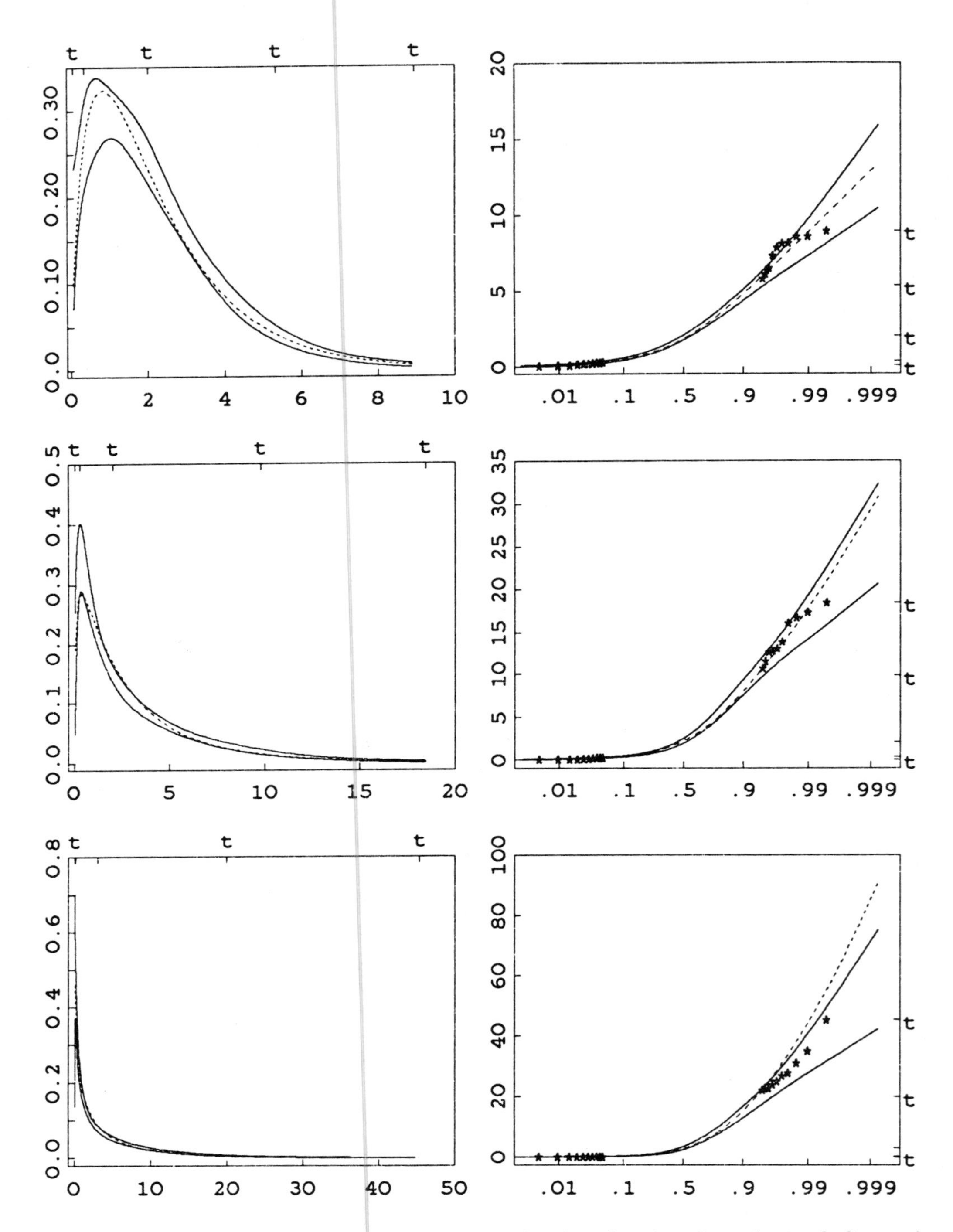

FIGURE 1b. 80% pointwise confidence intervals for density functions, left, and quantiles, right. From top to bottom $\beta \doteq 1.15,\ 1.52,\ 2$.

REFERENCES

1. C. de Boor, A Practical Guide to Splines, Springer-Verlag, New York, 1978.

2. L. Breiman and C. J. Stone, "Broad spectrum estimates and confidence intervals for tail quantiles", Tech. Rpt. No. 46, Dept. of Statist., Univ. of California, Berkeley, 1985.

3. N. L. JOHNSON, "Approximations to distributions," International Encyclopedia of Statistics, W. H. Kruskal and J. M. Tanur, Eds., The Free Press, New York, 1978, 169-174.

4. T. LEONARD, "Density estimation, stochastic processes and prior information" (with discussion), J. Roy. Statist. Soc. Ser. B, 40 (1978), 113-146.

5. B. W. SILVERMAN, "On the estimation of a probability density function by the maximum penalized likelihood method," Ann. Statist., 10 (1982), 795-810.

6. C. J. STONE, "Asymptotic properties of logspline density estimation," manuscript in preparation, 1986.

DEPARTMENT OF STATISTICS
UNIVERSITY OF CALIFORNIA, BERKELEY
BERKELEY, CALIFORNIA 94720

Contemporary Mathematics
Volume 59, 1986

Statistical Encounters with B-Splines

Wolfgang Dahmen and Charles A. Micchelli

1. Introduction.

Statistics and spline approximation are subjects having separate directions and adherents. However, as we know there is a fertile common ground, a small part of which provides the subject of this paper, [26]. We are concerned here with the appearance of the B-spline in all of its manifestations as a statistical entity.

The univariate B-spline with equally spaced knots was introduced into approximation theory by I.J. Schoenberg in a seminal paper on smoothing of data [22]. Schoenberg used Fourier analysis for his study of B-splines. Shorty afterwards in an announcement [4], H.B. Curry and Schoenberg suggested the use of divided differences as a way to define B-splines for an arbitrary knot sequence but the details of their results appeared quite a bit later [5]. B-splines are now an indispensable part of the theory and myriad applications of spline functions, [2,24].

The theory of multivariate splines is significantly more complex and only recently has become a focus of attention [7]. And so, it was no small accomplishment when more than twenty years ago Schoenberg, in a letter to P. Davis, gave a construction of a bivariate extension of the univariate B-spline. The basis of his observation was a geometric interpretation of the univariate B-spline given in [5]. There the Hermite-Genocchi formula for divided differences was used to show that the B-spline is the volume of the intersection of a hyperplane with the standard simplex. Since the

0271-4132/86 $1.00 + $.25 per page

Hermite-Genocchi formula also holds in the plane for analytic functions (which was the essence of Schoenberg's remark to P. Davis) bivariate B-splines can be defined by following the ideas of [5].

Using this geometric construction, de Boor introduced the **multivariate B-spline** at the end of his survey article [3]. This was an important step in the development of multivariate splines. However, there was no immediate reaction because it was difficult to make use of this definition. Now, we know much about the multivariate B-spline and also its offspring-**polyhedral B-splines**, [7]. More effort is needed, especially directed towards computation and applications, for multivariate splines to become as familiar and important as univariate splines.

One of our main reasons for writing this article was the surprising realization (at least to us) that the multivariate B-spline had an early presence in the statistical literature. In several papers on serial correlation coefficients [1,17,19-21,25], especially in the work of G.S. Watson [25], the multivariate B-spline makes a strong appearance. The univariate B-spline appeared in von Neumann's paper [19] and the special case discussed by Schoenberg [22] can be traced back to Laplace [18].

The emphasis of these papers is on the study of the joint density of certain random variables of statistical importance. However, even when these functions were found explicitly there seemed to be no importance placed on their smooth piecewise polynomial structure. This property is essential in approximation theory.

In this paper, we describe certain statistical encounters with B-splines. Our discussion is bound to be quite elementary to those well-versed in approximation theory and likewise parts will be quite trivial to statisticians. Nevertheless, we hope some value will be found in what we present here.

We begin in the second section with definitions and properties of univariate and multivariate B-splines as they relate to the joint density of serial correlation coefficients. In the third section we describe how B-splines arise in order statistics, [12,13] and the last section contains a description of an urn model which leads to B-splines. This work has applications in computer aided design (CAD) for the display of curves and surfaces, [10].

2. The B-spline.

Let $V_1, \ldots, V_n$ be independent random variables uniformly distributed over (0,1). Then the density M of $V_1 + \cdots + V_n$ is obtained by n-fold convolution. Therefore, the characteristic function of M is given by

$$E(e^{ix\bullet}) = \int_{-\infty}^{\infty} e^{itx} M(t)dt = \left(\int_0^1 e^{itx}dt\right)^n = \left(\frac{e^{ix}-1}{ix}\right)^n. \tag{2.1}$$

To obtain an explicit formula for M we use facts about divided differences. For this purpose, we let $\Delta g(t) = g(t+1) - g(t)$ be the forward difference operator. Then the n-th forward difference

$$\Delta^n g(0) = \sum_{k=0}^{n} \binom{n}{k} (-1)^{n-k} g(k) \tag{2.2}$$

is zero for any g which is a polynomial of degree $\leq n-1$. Hence, if we take any $g \in C^n(0, n)$ and expand it in a Maclaurin series with remainder we get

$$\Delta^n g(0) = \int_0^n \hat{M}(t) g^{(n)}(t)dt \tag{2.3}$$

where

$$\hat{M}(t) = \frac{1}{(n-1)!} \sum_{k=0}^{n} \binom{n}{k} (-1)^{n-k} (k-t)_+^{n-1} \tag{2.4}$$

and $t_+^{n-1} = (\max(t, 0))^{n-1}$. The fact that $\hat{M}$ is the density of the random variable $V_1 + \cdots + V_n$, that is, $M = \hat{M}$ follows from (2.3) by choosing $g(t) = e^{itx}$ and comparing to (2.1). Formula (2.3) is referred to as the Peano kernel representation of the divided difference. Formula (2.4) has been around for a long time. It was apparently known to Laplace [18].

From formula (2.3), (2.4) we see that $M(t)$ is a polynomial of degree $n - 1$ on each interval between integers, vanishes outside of $(0,n)$ and has $n - 2$ continuous derivatives everywhere. These properties determine M uniquely. To see this, we observe that any function with these properties must be a linear combination of the truncated powers, $t_+^{n-1}, (t-1)_+^{n-1}, \ldots, (t-n)_+^{n-1}$. For $t > n$ this linear combination must be zero. This results in a set of linear equations for the coefficients which is of full rank. Thus M is unique since its integral is one. Schoenberg calls M the **forward B-spline** with knots at $0, 1, \ldots, n$.

Next, following [5] we introduce the B-spline corresponding to an arbitrary knot sequence. For $t_0 < t_1 < \cdots < t_n$ the n-th divided difference of g at $t_0, \ldots, t_n$ is given by

$$[t_0, \ldots, t_n]\, g = \sum_{j=0}^{n} \frac{g(t_j)}{\prod_{\ell \neq j}(t_j - t_\ell)}. \tag{2.5}$$

Thus we have as before

$$[t_0, \ldots, t_n]\, g = \frac{1}{n!} \int M(t \mid t_0, \ldots, t_n) g^{(n)}(t) dt \tag{2.6}$$

where

$$M(t \mid t_0, \ldots, t_n) = n \sum_{j=0}^{n} \frac{(t_j - t)_+^{n-1}}{\prod_{\ell \neq j}(t_j - t_\ell)}. \tag{2.7}$$

This function is the **univariate B-spline** of degree $n - 1$ with knots at $t_0, t_1, \ldots, t_n$ normalized to have integral one. It has the following statistical interpretation, [8,16].

Let $\mathbf{V} = (V_0, \ldots, V_n)$ be a random vector uniformly distributed over the **standard n-simplex**

$$\Delta_n = \{(v_0, \ldots, v_n): v_j \geq 0, j = 0, 1, \ldots, n, \sum_{k=0}^{n} v_k = 1\}. \tag{2.8}$$

The expected value of the random variable $g(Z)$, $Z = t_0V_0 + \cdots + t_nV_n$ is given by

$$E(g(Z)) = n! \int_{\Delta_n} g(t_0v_0 + \cdots + t_nv_n)dv_1 \ldots dv_n. \tag{2.9}$$

To determine the density of Z we recall the Hermite-Genocchi formula

$$[t_0, \ldots, t_n]g = \int_{\Delta_n} g^{(n)}(t_0v_0 + \cdots + t_nv_n)dv_1 \ldots dv_n. \tag{2.10}$$

The validity of this formula can be established by induction and the recurrence relation for divided differences

$$[t_0, \ldots, t_n]g = \frac{[t_0, \ldots, t_{n-1}]g - [t_1, \ldots, t_n]g}{t_0 - t_n}. \tag{2.11}$$

Therefore, (2.6), (2.9) and (2.10) show that the density of $Z = t_0v_0 + \cdots + v_n$ is $\mathbf{M}(t \mid t_0, \ldots, t_n)$, since

$$\int_{-\infty}^{\infty} M(t \mid t_0, \ldots, t_n)g(t)dt = n! \int_{\Delta_n} g(t_0v_0 + \cdots + t_nv_n)dv_1 \ldots dv_n. \tag{2.12}$$

With this formula we can follow [5] and obtain a **geometric** interpretation of the B-spline. We let $\mathbf{y}^0, \dots, \mathbf{y}^n$ be **any** vectors in $\mathbb{R}^n$ such that

$$\mathbf{y}^i \mid_{\mathbb{R}^1} = t_i, \; i = 0, 1, \dots, n, \tag{2.13}$$

and

$$\mathrm{vol}_n \, [\mathbf{y}^0, \dots, \mathbf{y}^n] > 0 \tag{2.14}$$

where $[\mathbf{y}^0, \dots, \mathbf{y}^n]$ = convex hull $\{\mathbf{y}^0, \dots, \mathbf{y}^n\}$, that is, the simplex determined by $\mathbf{y}^0, \dots, \mathbf{y}^n$. Then by a change of variables (2.12) gives

$$M(t \mid t_0, \dots, t_n) = n! \frac{\mathrm{vol}_{n-1}\{\mathbf{y} \in [\mathbf{y}^0, \dots, \mathbf{y}^n] \colon \mathbf{y} \mid_{\mathbb{R}^1} = t\}}{\mathrm{vol}_n[\mathbf{y}^0, \dots, \mathbf{y}^n]} . \tag{2.15}$$

This formula was used by de Boor to define the **multivariate B-spline** in the following way. Let $\mathbf{x}^0, \dots, \mathbf{x}^n \in \mathbb{R}^s$, such that $\mathrm{vol}_s[\mathbf{x}^0, \dots, \mathbf{x}^n] > 0$ and suppose $\mathbf{y}^0, \dots, \mathbf{y}^n$ are **any** vectors satisfying (2.14) and

$$\mathbf{y}^i \mid_{\mathbb{R}^s} = \mathbf{x}^i, \quad i = 0, 1, \dots, n. \tag{2.16}$$

Then the multivariate B-spline is defined as

$$M(\mathbf{x} \mid \mathbf{x}^0, \dots, \mathbf{x}^n) = n! \frac{\mathrm{vol}_{n-s}\{\mathbf{y} \in [\mathbf{y}^0, \dots, \mathbf{y}^n] \colon \mathbf{y} \mid_{\mathbb{R}^s} = \mathbf{x}\}}{\mathrm{vol}_n[\mathbf{y}^0, \dots, \mathbf{y}^n]}. \tag{2.17}$$

Is the right hand side of (2.17) independent of **any** choice of vectors satisfying (2.14), (2.16) as in the univariate case? Yes, because by reversing the above reasoning the multivariate B-spline can also be defined by the formula

$$\int_{\mathbb{R}^s} f(\mathbf{x}) M(\mathbf{x} \mid \mathbf{x}^0, \dots, \mathbf{x}^n) d\mathbf{x} = n! \int_{\Delta_n} f(v_0 \mathbf{x}^0 + \cdots + v_n \mathbf{x}^n) dv_1 \dots dv_n. \tag{2.18}$$

Thus $M(\mathbf{x} \mid \mathbf{x}^0, \ldots, \mathbf{x}^n)$ is the joint density of the random vector $\mathbf{Z} = \mathbf{x}^0 V_0 + \cdots + \mathbf{x}^n V_n$ and for $s = 1$, it reduces to the univariate B-spline. In general, it has many properties similar to the univariate B-spline. For instance, it is a piecewise polynomial of degree $\leq n - s$ with compact support on $[\mathbf{x}^0, \ldots, \mathbf{x}^n]$ which has $n - s - 1$ continuous derivatives when $\mathbf{x}^0, \ldots, \mathbf{x}^n$ are in general position, [7].

We will give a formula for $M(\mathbf{x} \mid \mathbf{x}^0, \ldots, \mathbf{x}^n)$ but first we follow [16] and consider the following general construction. Suppose the random vector $\mathbf{V} = (V_1, \ldots, V_n)$ has a joint probability density $\varphi(\mathbf{v})$ and $\mathbf{x}^1, \ldots, \mathbf{x}^n$ are vectors which span $\mathbb{R}^s$. Then the random vector $\mathbf{Z} = V_1\mathbf{x}^1 + \cdots + V_n\mathbf{x}^n$ has a joint density, $N_\varphi(x \mid \mathbf{x}^1, \ldots, \mathbf{x}^n)$, $\mathbf{x} \in \mathbb{R}^s$ defined by

$$\int_{\mathbb{R}^s} f(\mathbf{x}) N_\varphi(\mathbf{x} \mid \mathbf{x}^1, \ldots, \mathbf{x}^n) d\mathbf{x} = \int_{\mathbb{R}^n} f(v_1\mathbf{x}^1 + \cdots + v_n\mathbf{x}^n)\varphi(v_1, \ldots, v_n) dv_1 \ldots dv_n. \quad (2.19)$$

In the special case when $\mathbf{V}$ is uniformly distributed over Δ_n we have

$$M(\mathbf{x} \mid \mathbf{x}^0, \ldots, \mathbf{x}^n) = N_\varphi(\mathbf{x} + \mathbf{x}^0 \mid \mathbf{x}^1, \ldots, \mathbf{x}^n). \quad (2.20)$$

We also mention that whenever $\varphi(\mathbf{v})$ is a piecewise polynomial over some triangulation of a bounded domain in $\mathbb{R}^n$, $N_\varphi(\mathbf{x} \mid \mathbf{x}^1, \ldots, \mathbf{x}^n)$ is also a piecewise polynomial on $\mathbb{R}^s$.

There are some important special cases of (2.19) which will come up in our later remarks. Let C be a polyhedral subset of $\mathbb{R}^s$ and $\mathbf{V}$ a random vector uniformly distributed over C. In this case we call the density (2.19) a **polyhedral B-spline.** When $C = [0, 1]^n$ we get the **box spline** which we denote by $B(\mathbf{x} \mid \mathbf{x}^1, \ldots, \mathbf{x}^n)$ and so it is defined by the equation

$$\int_{\mathbb{R}^s} f(\mathbf{x}) B(\mathbf{x} \mid \mathbf{x}^1, \ldots, \mathbf{x}^n) d\mathbf{x} = \int_0^1 \cdots \int_0^1 f(v_1\mathbf{x}^1 + \cdots + v_n\mathbf{x}^n) dv_1 \ldots dv_n. \quad (2.21)$$

When $C = \mathbb{R}^n_+$ and $0 \notin [\mathbf{x}^1, \ldots, \mathbf{x}^n]$ we get the **multivariate truncated power** which we denote by $T(\mathbf{x} \mid \mathbf{x}^1, \ldots, \mathbf{x}^n)$. It is defined by the equation

$$\int_{\mathbb{R}^s} f(\mathbf{x})T(\mathbf{x} \mid \mathbf{x}^1, \ldots, \mathbf{x}^n)d\mathbf{x} = \int_0^\infty \cdots \int_0^\infty f(v_1\mathbf{x}^1 + \cdots + v_n\mathbf{x}^n)dv_1 \ldots dv_n. \tag{2.22}$$

For $s = 1$, the truncated power can be computed explicitly

$$T(t \mid x_1, \ldots, x_n) = \frac{1}{(n-1)!} \frac{t_+^{n-1}}{x_1 \ldots x_n}. \tag{2.23}$$

A formula for the univariate box spline is given later.

Finally, we mention the **generalized Dirichlet density**

$$\int_{\mathbb{R}^s} f(\mathbf{x})D_\alpha(\mathbf{x} \mid \mathbf{x}^0, \ldots, \mathbf{x}^n)d\mathbf{x} = \int_{\Delta_n} f(v_0\mathbf{x}^0 + \cdots + v_n\mathbf{x}^n)D_\alpha(\mathbf{v})dv_1 \ldots dv_n \tag{2.24}$$

where

$$D_\alpha(\mathbf{v}) = \frac{\Gamma(\alpha_0 + \cdots + \alpha_n)}{\Gamma(\alpha_0) \ldots \Gamma(\alpha_n)} \prod_{i=0}^n v_i^{\alpha_i - 1}, \mathbf{v} \in \Delta_n, \alpha_0 > 0, \ldots, \alpha_n > 0,$$

and the **generalized Gamma density**

$$\int_{\mathbb{R}^s} f(\mathbf{x})G_\alpha(\mathbf{x} \mid \mathbf{x}^1, \ldots, \mathbf{x}^n)d\mathbf{x} = \int_{\mathbb{R}_+^s} f(v_1\mathbf{x}^1 + \cdots + v_n\mathbf{x}^n)G_\alpha(\mathbf{v})dv_1 \ldots dv_n, \tag{2.25}$$

where

$$G_\alpha(\mathbf{v}) = \frac{1}{\Gamma(\alpha_1) \ldots \Gamma(\alpha_n)} e^{-(v_1 + \cdots + v_n)} \prod_{i=1}^n v_i^{\alpha_i - 1}, \mathbf{v} \in \mathbb{R}_+^n,$$

both of which appear in [16]. Some of these special densities will be useful in our discussion of the joint density for **serial correlation coefficients.**

For this purpose we consider independent identically distributed random variables $U_1, \ldots, U_N$ having a normal distribution with mean zero and variance one. Let $A_1, \ldots, A_s$ be s nonsingular $N \times N$ real symmetric matrices and consider the random vector $\mathbf{R} = (R_1, \ldots, R_s)$

$$R_i = \frac{(A_i\mathbf{U}, \mathbf{U})}{(\mathbf{U}, \mathbf{U})}, \quad i = 1, \ldots, s, \ \mathbf{U} = (U_1, \ldots, U_N). \tag{2.26}$$

We remark that since R_i is unchanged if all the variables are scaled by the same positive number no greater generally is obtained if each U_i have the same variance different from one. We will show that under certain circumstances the joint density of $\mathbf{R} = (R_1, \ldots, R_s)$ is a **multivariate B-spline.** To this end, we require that $A_1, \ldots, A_s$ are simultaneously diagonalizable. Thus,

$$A_i = O^T D_i O, \quad i = 1, 2, \ldots, s \tag{2.27}$$

for some orthogonal matrix O and diagonal matrices $D_i = \operatorname{diag}(\lambda_1^i, \ldots, \lambda_N^i)$. Setting $\mathbf{Y} = \mathbf{OU}$ then $\mathbf{Y}$ is a normal with mean zero and covariance I and

$$R_i = \sum_{\ell=1}^{N} \lambda_\ell^i U_\ell^2 \Big/ \sum_{\ell=1}^{N} U_\ell^2, \quad i = 1, \ldots, s. \tag{2.28}$$

We mention two examples of this problem.

Example 2.1. von Neumann [20] considered the random variable $R = \delta^2/s^2$ where

$$\delta^2 = \frac{1}{N-1} \sum_{\mu=1}^{N-1} (U_{\mu+1} - U_\mu)^2$$

$$s^2 = \frac{1}{N} \sum_{\mu=1}^{N} (U_\mu - \overline{U})^2, \ \overline{U} = \frac{1}{N} \sum_{\mu=1}^{N} U_\mu$$

and each U_i is normal with mean ξ and standard deviation σ. Thus we may express R as a ratio of two quadratic forms $R = (A\,U, U)/(B\,U, U)$. The matrix B is given by $B = I - N^{-1}1$ where 1 is the matrix all of whose entries are one and the matrix in the numerator of R is

$$A = \frac{N}{N-1}\begin{bmatrix} 1 & -1 & & & & & \\ -1 & 2 & -1 & & & & \\ & -1 & 2 & -1 & & & \\ & & \cdot & \cdot & \cdot & & \\ & & & \cdot & \cdot & \cdot & \\ & & & & -1 & 2 & -1 \\ & & & & & -1 & 1 \end{bmatrix}.$$

These matrices commute and A has eigenvalues $4\sin^2\frac{k\pi}{N}$, $k = 0, 1, \ldots, N-1$ while the eigenvalues of B are 0 and 1, the latter having multiplicity $N - 1$. Thus in this case R reduces to (2.28) with $s = 1$, N replaced by $N - 1$ and $\lambda_\ell = 4\sin^2\frac{\ell\pi}{N}$, $\ell = 1, \ldots, N-1$. Furthermore, setting $R = \frac{2N}{N-1}(1-\varepsilon)$, as in [20], gives

$$\varepsilon = \frac{\sum_{\mu=1}^{N-1} \cos\frac{\mu\pi}{N} Y_\mu^2}{\sum_{\mu=1}^{N-1} Y_\mu^2}. \tag{2.29}$$

Example 2.2. In [20], Quenouille found the joint density for the serial correlation coefficients

$$R_\ell = \frac{\sum_{i=1}^{N} V_i V_{i+\ell} - \left(\sum_{i=1}^{N} V_i\right)^2 / N}{\sum_{i=1}^{N} V_i^2 - \left(\sum_{i=1}^{N} V_i\right)^2 / N} = \frac{\sum_{i=1}^{N} (V_i - \overline{V})(V_{i+\ell} - \overline{V})}{\sum_{i=1}^{N} (V_i - \overline{V})^2}, \quad \ell = 1, \ldots, s,$$

where V_i are normal and independently distributed with mean μ and variance σ^2 and $V_{N+i} = V_i$. The case of one correlation coefficient and any lag ℓ was considered by Anderson [1].

If we expand V_j in its finite Fourier series

$$V_j = \frac{1}{N} \sum_{k=1}^{N} \omega^{jk} \hat{V}_k, \quad \omega^N = 1,$$

and set $Y_k = | \hat{V}_k |^2$ we see that R_ℓ reduces to the form

$$R_\ell = \frac{\sum_{k=1}^{N-1} \cos \frac{2\pi \ell k}{N} Y_k^2}{\sum_{k=1}^{N-1} Y_k^2}. \tag{2.30}$$

As we will later see it is important that the eigenvalues here occur in pairs with the possible exception of -1 because

$$\cos \frac{2\pi j(N-k)}{N} = \cos \frac{2\pi jk}{N}. \tag{2.31}$$

This influences the final form of the joint distribution.

Returning to (2.28) we consider the situation when there are positive integers $p_0, \ldots, p_n$ such that for **any** i the eigenvalues of A_i are of the same multiplicity $p_0, \ldots, p_n$ and we let $x_i^0, \ldots, x_i^n$ be the corresponding distinct eigenvalues.

Hence

$$R_i = \sum_{j=0}^{n} x_i^j V_j, \quad i = 1, \ldots, s, \tag{2.32}$$

where

$$V_j = \sum_{\ell = q_j}^{q_{j+1}-1} Y_\ell^2 \Big/ \sum_{k=1}^{N} Y_k^2, \quad j = 0, 1, \ldots, n \tag{2.33}$$

and $q_0 = 0$, $q_j = p_0 + \cdots + p_{j-1}$, $j = 1, \ldots, n+1$. If we let $V_j = \Omega_j / \sum_{k=0}^{n} \Omega_n$ where

$$\Omega_j = \sum_{\ell = q_j}^{q_{j+1}-1} Y_\ell^2, \quad j = 0, \ldots, n,$$

then Ω_j has a chi square density with $p_j = q_{j+1} - q_j$ degrees of freedom, that is, its density is

$$\frac{1}{2^{p_j/2}\Gamma(p_j/2)} e^{-\omega_j/2} \omega_j^{p_j/2-1} . \tag{2.34}$$

Identifying the joint density of $\mathbf{V} = (V_0, \ldots, V_n)$ is a standard matter, Cf. Johnson and Kotz (1972), [14] where the following is proved.

If $\Omega_0, \ldots, \Omega_n$ are independent random variables such that each Ω_j is χ^2 with p_j degrees of freedom then $\mathbf{V} = (V_0, \ldots, V_n)$, $V_j = \Omega_j / \sum_{k=0}^{n} \Omega_k$ has a Dirichlet density with parameters $\alpha_j = p_j/2$,

$$D_\alpha(\mathbf{v}) = \frac{\Gamma(\alpha_0 + \cdots + \alpha_n)}{\Gamma(\alpha_0) \ldots \Gamma(\alpha_n)} v_0^{\alpha_0 - 1} \ldots v_n^{\alpha_n - 1}, \mathbf{v} \in \Delta_n. \tag{2.35}$$

This follows by noting that the conditional density of $(V_0, \ldots, V_n)$ given $\Omega_0 + \cdots + \Omega_n = \Omega$ is

$$\frac{1}{2^{\alpha_0 + \cdots + \alpha_n}\Gamma(\alpha_0 + \cdots + \alpha_n)} e^{-\frac{1}{2}\omega} \omega^{\alpha_0 + \cdots + \alpha_n - 1} D_\alpha(\mathbf{v})$$

and that the integral of this function over $\omega \in (0, \infty)$ is (2.35).

Thus we see that the joint density of $\mathbf{R}$ is a generalized Dirichlet density with parameters which are 1/2 of integers. Motivated by this fact we recall some properties of the generalized Dirichlet density.

When two vectors are repeated in the definition of the generalized Dirichlet density we may express the resulting function in two ways, specifically,

$$D_{(\alpha_0,\ldots,\alpha_n)}(\mathbf{x} \mid \overset{0}{\mathbf{x}}, \overset{0}{\mathbf{x}}, \overset{2}{\mathbf{x}}, \ldots, \overset{n}{\mathbf{x}}) = D_{(\alpha_0+\alpha_1,\alpha_2,\ldots,\alpha_n)}(\mathbf{x} \mid \overset{0}{\mathbf{x}}, \ldots, \overset{n-1}{\mathbf{x}}).$$

More generally, if $\mathbf{V} = (V_0, \ldots, V_n)$ has a Dirichlet density with parameters $\alpha_0, \ldots, \alpha_n$ then $\Omega = (\Omega_0, \ldots, \Omega_m)$ with

$$\Omega_\ell = \sum_{j=k_\ell}^{k_{\ell+1}-1} V_j, \quad \ell = 0, \ldots, m,$$

$k_0 = 0,\ k_{m+1} \le n+1$ has a Dirichlet density with parameters $\mu_0 = \alpha_0 + \cdots + \alpha_{k_1-1}, \ldots, \mu_m = \alpha_{k_m} + \cdots + \alpha_{k_{m+1}-1}$.

This shows that the multivariate B-spline with multiple knots corresponds to a Dirichlet density with integer parameters greater than one. Specifically, when only s + 1 of the knots are distinct, say $\overset{0}{\mathbf{x}}, \ldots, \overset{s}{\mathbf{x}}$, each repeated with multiplicities $\alpha_0, \ldots, \alpha_s$, we have

$$\begin{aligned} &M(\mathbf{x} \mid \underbrace{\overset{0}{\mathbf{x}}, \ldots, \overset{0}{\mathbf{x}}}_{\alpha_0}, \underbrace{\overset{1}{\mathbf{x}}, \ldots, \overset{1}{\mathbf{x}}}_{\alpha_1}, \ldots, \underbrace{\overset{s}{\mathbf{x}}, \ldots, \overset{s}{\mathbf{x}}}_{\alpha_s}) \\ &= D_\alpha(\lambda) / \det \begin{pmatrix} 1, 1, \ldots, 1 \\ \overset{0}{\mathbf{x}}, \overset{1}{\mathbf{x}}, \ldots, \overset{s}{\mathbf{x}} \end{pmatrix} \end{aligned} \tag{2.36}$$

where $\lambda = (\lambda_0, \ldots, \lambda_s)$ are the barycentric coordinates of $\mathbf{x}$ relative to the simplex generated by $\overset{0}{\mathbf{x}}, \ldots, \overset{s}{\mathbf{x}}$, that is,

$$\mathbf{x} = \sum_{i=0}^{s} \lambda_i \overset{i}{\mathbf{x}}, \quad \sum_{i=0}^{s} \lambda_i = 1.$$

In general we cannot find explicit formulas for the densities (2.24), (2.25), however, several useful formulas are available. For instance, by integrating the right hand side of (2.24) first for $v_1 + \cdots + v_n = t$ and then over $t \in (0,1)$ it can be shown that

$$D_\alpha(\mathbf{x} \mid \mathbf{x}^0, \ldots, \mathbf{x}^n) =$$

$$\frac{\Gamma(\alpha^0 + \cdots + \alpha^n)}{\Gamma(\alpha_1 + \cdots + \alpha_n)\,\Gamma(\alpha_0)} \int_1^\infty t^{-\sum_{j=0}^n \alpha_j + s}(t-1)^{\alpha_0 - 1} D_\alpha((1-t)\mathbf{x}^0 + t\mathbf{x} \mid \mathbf{x}^1, \ldots, \mathbf{x}^n)dt.$$

Also, as was pointed out in [16] the generalized Dirichlet density satisfies the equation

$$\int_{\mathbb{R}^s} (1 + \lambda \cdot \mathbf{x})^{-\sum_{i=0}^n \alpha_i} D_\alpha(\mathbf{x} \mid \mathbf{x}^0, \ldots, \mathbf{x}^n)d\mathbf{x} = \prod_{i=0}^n (1 + \lambda \cdot \mathbf{x}^i)^{-\alpha_i}. \tag{2.37}$$

To prove this equation, observe that the left hand side equals

$$\frac{1}{\Gamma(\alpha_0 + \cdots + \alpha_n)} \int_0^\infty t^{\sum_{i=0}^n \alpha_i - 1} e^{-t} \int_{\mathbb{R}^s} e^{-t\lambda \cdot \mathbf{x}} D_\alpha(\mathbf{x} \mid \mathbf{x}^0, \ldots, \mathbf{x}^n)d\mathbf{x}$$

which from definition (2.24) is the same as

$$= \frac{1}{\Gamma(\alpha_0) \ldots \Gamma(\alpha_n)} \int_0^\infty t^{\sum_{i=0}^n \alpha_i - 1} e^{-t} \int_{\Delta_n} e^{-tv_0\lambda \cdot \mathbf{x}^0 - \cdots - tv_n\lambda \cdot \mathbf{x}^n} v_0^{\alpha_0 - 1} \ldots v_n^{\alpha_n - 1} dv_1 \ldots dv_n dt$$

$$= \frac{1}{\Gamma(\alpha_0) \ldots \Gamma(\alpha_n)} \int_0^\infty e^{-t} \int_{t\Delta_n} e^{-tv_0\lambda \cdot \mathbf{x}^0 - \cdots - tv_n\lambda \cdot \mathbf{x}^n} v_0^{\alpha_0 - 1} \ldots v_n^{\alpha_n - 1} dv_1 \ldots dv_n dt$$

$$= \prod_{j=0}^n \left(\frac{1}{\Gamma(\alpha_j)} \int_0^\infty v_j^{\alpha_j - 1} e^{-v_j(1 + \lambda \cdot \mathbf{x}^j)} dv_j \right)$$

$$= \prod_{j=0}^n (1 + \lambda \cdot \mathbf{x}^j)^{-\alpha_j}.$$

Formula (2.37) is useful for obtaining explicit expressions for $D_\alpha(\mathbf{x} \mid \mathbf{x}^0, \ldots, \mathbf{x}^n)$. We will do this in two extreme cases, namely, $s = 1$, α_j arbitrary and s arbitrary, $\alpha_0 = \cdots = \alpha_n = 1$ and $0, \mathbf{x}^0, \ldots, \mathbf{x}^n$ are in general position. We will now deal with the first case and postpone the other case until later.

When $s = 1$, we can rewrite (2.37) in the equivalent form

$$\int_{\mathbb{R}} D_\alpha(t \mid t_0, \ldots, t_n)(\xi - t)^{-\beta} dt = \prod_{k=0}^{n} (\xi - t_k)^{-\alpha_k} \tag{2.38}$$

where $\beta = \alpha_0 + \cdots + \alpha_n$ and $\xi \in \mathbb{C}\backslash\mathbb{R}$. We assume $t_0 < \cdots < t_n$ and prove the formula

$$D_\alpha(t \mid t_0, \ldots, t_n) = \frac{\beta - 1}{2\pi i} \int_{\Gamma_{r+1}} \frac{(z - t)^{\beta-2}}{\prod_{k=0}^{n} (z - t_k)^{\alpha_k}} dz, \quad t_r < t < t_{r+1}, \tag{2.39}$$

$r = 0, 1, \ldots, n - 1$, where Γ_{r+1} is any contour enclosing $t_{r+1}, \ldots, t_n$ but not $t_0, \ldots, t_r$. Here we choose the branch of the integrand so that it is positive for z positive and greater than t_0. A special case of this formula was established by Koopmans [17].

We prove (2.39) under the condition $\beta \geq \alpha_j + 1$, $j = 0, 1, \ldots, n - 1$, which is sufficient for the special cases we have in mind. First note that the integral on the right of (2.39), which we call $g(t)$ is zero for $t < t_0$ because the contour includes all the singularities of the denominator and the integrand decays like $0(\mid z \mid^{-2})$, $z \to \infty$. For $t > t_n$, $g(t)$ is zero because the integrand is analytic within the contour. Therefore for $\xi \in \mathbb{C}\backslash\mathbb{R}$.

$$\int_{\mathbb{R}} g(t) \frac{dt}{(\xi - t)^{\beta}} = \sum_{j=0}^{n-1} \int_{t_j}^{t_{j+1}} g(t) \frac{dt}{(\xi - t)^{\beta}}$$

$$= \sum_{j=0}^{n-1} \frac{1}{2\pi i} \int_{\Gamma_{j+1}} \frac{1}{\prod_{k=0}^{n} (z-t_k)^{\alpha_k}} (\beta-1) \int_{t_j}^{t_{j+1}} \left(\frac{z-t}{\xi-t}\right)^{\beta-2} \frac{dt}{(\xi-t)^2}$$

$$= \sum_{j=0}^{n-1} \frac{1}{2\pi i} \int_{\Gamma_{j+1}} \frac{1}{\prod_{k=0}^{n} (z-t_k)^{\alpha_k}} \frac{1}{z-\xi} \left[\left(\frac{z-t_j}{\xi-t_j}\right)^{\beta-1} - \left(\frac{z-t_{j+1}}{\xi-t_{j+1}}\right)^{\beta-1}\right] dz.$$

We choose all our contours so that ξ is in their exterior. Since $\beta \geq \alpha_j + 1$, the first integrand in the summation above is analytic in a neighborhood of t_j, $j = 0, 1, \ldots, n-1$. Hence we may move the contour Γ_{j+1} to Γ_j in this integral. Thus we obtain a telescoping sum which reduces to

$$\frac{1}{2\pi i} \int_{\Gamma_0} \frac{1}{z-\xi} \frac{\left(\frac{z-t_0}{\xi-t_0}\right)^{\beta-1}}{\prod_{k=0}^{n} (z-t_k)^{\alpha_k}} dz$$

which can be evaluated because, again, the integrand decays as $0(|z|^{-2})$ as $z \to \infty$. Thus the sum of all the residues is zero and since the residue at $z = \xi$ is

$$\frac{1}{\prod_{k=0}^{n} (\xi - t_k)^{\alpha_k}}$$

formula (2.39) is verified. Here we used the fact that (2.38) determines $D_\alpha(t \mid t_0, \ldots, t_n)$ uniquely. When $\alpha_0 = \cdots = \alpha_n = 1$ then (2.39) reduces to the truncated power representation (2.7) for the B-spline.

For $\alpha_0 = \frac{1}{2}$, $\alpha_1 = \cdots = \alpha_n = 1$ (2.39) gives

$$D_\alpha(t \mid t_0, \ldots, t_n) = (n - 1/2) \sum_{k=r+1}^{n} \frac{(t_k - t)^{n-3/2}}{\prod_{j \neq k} (t_k - t_j)\sqrt{t_k - t_0}}, \quad t_r < t < t_{r+1}.$$

This function arises as the density of the serial correlation coefficient

$$\frac{x_1 x_{\ell+1} + \cdots + x_N x_\ell - (\Sigma x_i)^2/N}{\Sigma x_i^2 - (\Sigma x_i)^2/N}$$

when $\ell = 1$, N is even, $N = 2n + 2$ see [1]. In this case, according to Example 2.2. the eigenvalues are $\cos \frac{2\pi k}{N}$, $k = 1, \ldots, N - 1$. Each has multiplicity two except for $k = n + 1$ which is a simple eigenvalue. Thus $t_k = -\cos \frac{\pi k}{n+1}$, $k = 0, 1, \ldots, n$ gives the proper density in this case.

A contour integral formula for the univariate generalized Gamma density can also be derived. We state the following without proof the formula.

$$\Lambda_\alpha(t \mid t_0, \ldots, t_n) = \begin{cases} \dfrac{1}{2\pi i} \displaystyle\int_{\Gamma_+} \frac{e^{-tz}}{\prod_{i=0}^{n} (1 - zt_i)^{\alpha_i}} \, dz, & t \geq 0 \\ \dfrac{1}{2\pi i} \displaystyle\int_{\Gamma_-} \frac{e^{-tz}}{\prod_{i=0}^{n} (1 - zt_i)^{\alpha_i}} \, dz, & t \leq 0 \end{cases}$$

where Γ_+, Γ_- are contours in the right half, left half plane containing the poles of the integrand which are positive, negative, respectively.

This formula is obtained by Fourier inversion followed by an appropriate modification of the contour away from the imaginary axis.

Let us now turn our attention to the derivation of explicit representations for the multivariate Dirichlet and Gamma densities with parameters $\alpha_j = 1$ for all j. This will lead us as well to a formula for the multivariate truncated power, see (2.22).

To this end, let us suppose throughout the following that $0, x^0, \ldots, x^n \in \mathbb{R}^s$ are in general position. That is, every subset of $s+1$ of these vectors form a proper s-dimensional simplex. Specializing (2.37) to the case $\alpha_j = 1, j = 0, \ldots, n$, yields

$$\int_{\mathbb{R}^s} (1 + \lambda \cdot \mathbf{x})^{-n-1} M(\mathbf{x} \mid \mathbf{x}^0, \ldots, \mathbf{x}^n) d\mathbf{x} = 1 / \prod_{j=0}^{n} (1 + \lambda \cdot \mathbf{x}^j). \tag{2.40}$$

Our goal is a partial fraction decomposition of the right hand side of (2.40). Note that under the above assumptions on $\mathbf{x}^0, \ldots, \mathbf{x}^n$ there exists for each $I = \{i^1, \ldots, i^s\} \subset \{0, \ldots, n\}$ a unique $\mathbf{x}^I \in \mathbb{R}^s$ such that

$$1 + \mathbf{x}^I \cdot \mathbf{x}^i = 0, \quad i \in I,$$

while $1 + \mathbf{x}^I \cdot \mathbf{x}^j \neq 0$, $j \notin I$. Hence defining

$$Q_I(\mathbf{x}) = \prod_{j \notin I} \frac{(1 + \mathbf{x} \cdot \mathbf{x}^j)}{(1 + \mathbf{x}^j \cdot \mathbf{x}^I)}$$

we have $Q_I(\mathbf{x}^{I'}) = \delta_{I, I'}$. Thus each polynomial P of degree $\leq n - s + 1$ can be written as

$$P(\mathbf{x}) = \sum_{|I| = s} P(\mathbf{x}^I)\, Q_I(\mathbf{x}).$$

Specifically, choosing $P = 1$ we obtain

$$\frac{1}{\prod_{j=0}^{n}(1+\lambda \cdot \mathbf{x}^{j})} = \sum_{|I|=s} a_I \Big/ \prod_{j=1}^{s}(1+\lambda \cdot \mathbf{x}^{i_j}) \tag{2.41}$$

where

$$a_I = 1 \Big/ \prod_{j \notin I}(1+\mathbf{x}^{I} \cdot \mathbf{x}^{j}) = \frac{(\det(\mathbf{x}^{i_1}, \dots, \mathbf{x}^{i_s}))^{n-s+1}}{\prod_{j\notin I} \det\begin{pmatrix} 1, & 1, & \dots & , 1 \\ \mathbf{x}^{j}, & \mathbf{x}^{i_1}, & \dots & , \mathbf{x}^{i_s} \end{pmatrix}} . \tag{2.42}$$

Inserting (2.41) into (2.40) yields

$$\int_{\mathbb{R}^s} (1+\lambda \cdot \mathbf{x})^{-n-1} \left(M(\mathbf{x} \mid \mathbf{x}^{0}, \dots, \mathbf{x}^{n}) - \sum_{|I|=s} a_I M(\mathbf{x} \mid \underbrace{0, \dots, 0}_{n-s+1}, \mathbf{x}^{i_1}, \dots, \mathbf{x}^{i_s}) \right) d\mathbf{x} = 0.$$

Appealing to the denseness of the family $\{(1+\lambda \cdot \mathbf{x})^{-n-1} : \lambda \in \mathbb{R}^s\}$ with respect to uniform convergence on compact subsets in $\mathbb{R}^s$ we conclude that

$$M(\mathbf{x} \mid \mathbf{x}^{0}, \dots, \mathbf{x}^{n}) = \sum_{|I|=s} a_I M(\mathbf{x} \mid \underbrace{0, \dots, 0}_{n-s-1}, \mathbf{x}^{i_1}, \dots, \mathbf{x}^{i_s}) . \tag{2.43}$$

Each multivariate B-spline on the right hand side has only $s+1$ distinct vectors and is therefore known to agree on its support with a Bernstein polynomial, see (2.36) when $\alpha_0 = \cdots = \alpha_n = 1$. More precisely, we have for $\sigma_I = [0, \mathbf{x}^{i_1}, \dots, \mathbf{x}^{i_s}]$

$$M(\mathbf{x} \mid 0, \dots, 0, \mathbf{x}^{i_1}, \dots, \mathbf{x}^{i_s}) =$$

$$s! \binom{n}{s} \left(\frac{\det\begin{pmatrix} 1, & 1, & \dots, & 1 \\ \mathbf{x}, & \mathbf{x}^{i_1}, & \dots & , \mathbf{x}^{i_s} \end{pmatrix}}{\det(\mathbf{x}^{i_1}, \dots, \mathbf{x}^{i_s})} \right)^{n-s} \frac{\operatorname{sign}(\det(\mathbf{x}^{i_1}, \dots, \mathbf{x}^{i_s}))}{\det(\mathbf{x}^{i_1}, \dots, \mathbf{x}^{i_s})} \chi_{\sigma_I}(\mathbf{x}) . \tag{2.44}$$

Combining (2.42), (2.43) and (2.44) provides

$$M(\mathbf{x} \mid \mathbf{x}^0, \ldots, \mathbf{x}^n) = \frac{n!}{(n-s)!} \sum_{|I|=s} \frac{\chi_{\sigma_I}(\mathbf{x}) \operatorname{sgn}(\det(\mathbf{x}^{i_1}, \ldots, \mathbf{x}^{i_s})) \left(\det \begin{pmatrix} 1, & 1, & \ldots, & 1 \\ \mathbf{x}, & \mathbf{x}^{i_1}, & \ldots, & \mathbf{x}^{i_s} \end{pmatrix} \right)^{n-s}}{\prod_{j \notin I} \det \begin{pmatrix} 1, & 1, & \ldots, & 1 \\ \mathbf{x}^j, & \mathbf{x}^{i_1}, & \ldots, & \mathbf{x}^{i_s} \end{pmatrix}} \tag{2.45}$$

$$= \frac{n!}{(n-s)!} \sum_{|I|=s} \frac{\chi_{\sigma_I}(\mathbf{x})}{|\det(\mathbf{x}^{i_1}, \ldots, \mathbf{x}^{i_s})|} \frac{(1 + x^I \cdot \mathbf{x})^{n-s}}{\prod_{j \notin I} (1 + x^I \cdot \mathbf{x}^j)} .$$

$$= \frac{n!}{(n-s)!} \sum_{|I|=s} \frac{\chi_{\sigma_I}(\mathbf{x})}{|\det(\mathbf{x}^{i_1}, \ldots, \mathbf{x}^{i_s})|} a_I (1 + \mathbf{x} \cdot \mathbf{x}^I)^{n-s}$$

This identity was first proved by Watson [25] and later in Hu and Lee [11] by different methods.

The argument for the Gamma density with coefficients $\alpha_j = 1$, $j = 1, \ldots, n$, is similar. Here we exploit the fact that

$$\int_{\mathbb{R}^s} e^{\lambda \cdot \mathbf{x}} G(\mathbf{x} \mid \mathbf{x}^1, \ldots, \mathbf{x}^n) d\mathbf{x} = 1 / \prod_{j=1}^{n} (1 + \lambda \cdot \mathbf{x}^j)$$

to conclude just as above that

$$G(\mathbf{x} \mid \mathbf{x}^1, \ldots, \mathbf{x}^s) = \sum_{|I|=s} a_I \, G(\mathbf{x} \mid \mathbf{x}^{i_1} \ldots \mathbf{x}^{i_s}) . \tag{2.46}$$

Setting

$$Y_I = \{\mathbf{x}^{i_1}, \ldots, \mathbf{x}^{i_s}\}$$

one easily confirms that

$$G(\mathbf{x} \mid \mathbf{x}^{i_1}, \ldots, \mathbf{x}^{i_s}) = \mid \det Y_I \mid^{-1} e^{\mathbf{x}^I \bullet \mathbf{x}} \chi_{<I>_+}(\mathbf{x})$$

where $< I >_+$ denotes the cone spanned by Y_I. This gives

$$G(\mathbf{x} \mid \mathbf{x}^1, \ldots, \mathbf{x}^n) = \sum_{|I|=s} \frac{1}{\prod_{j \notin I} (1 + \mathbf{x}^I \bullet \mathbf{x}^j)} \mid \det Y_I \mid^{-1} e^{\mathbf{x}^I \bullet \mathbf{x}} \chi_{<I>_+}(\mathbf{x}). \tag{2.47}$$

In order to derive an explicit representation for the truncated power $T(\bullet \mid \mathbf{x}^1, \ldots, \mathbf{x}^n)$, defined by (2.22) when $0 \notin [\mathbf{x}^1, \ldots, \mathbf{x}^n]$ we define for $\mu > 0$ another density $\Lambda^{(\mu)}(\bullet \mid \mathbf{x}^1, \ldots, \mathbf{x}^n)$ by requiring that

$$\int_{\mathbb{R}^s} \Lambda^{(\mu)}(\mathbf{x} \mid \mathbf{x}^1, \ldots, \mathbf{x}^n) f(\mathbf{x}) d\mathbf{x} = \int_{\mathbb{R}^n_+} e^{-\mu(t_1 + \cdots + t_n)} f(t_1\mathbf{x}^1 + \cdots + t_n\mathbf{x}^n) \, dt_1 \ldots dt_n.$$

Thus

$$\Lambda^{(\mu)}(\bullet \mid \mathbf{x}^1, \ldots, \mathbf{x}^n) = \mu^{-n} G(\bullet \mid \mathbf{x}^1/\mu, \ldots, \mathbf{x}^n/\mu) \tag{2.48}$$

and $\Lambda^{(\mu)}(\bullet \mid \mathbf{x}^1, \ldots, \mathbf{x}^n)$ tends to $T(\bullet \mid \mathbf{x}^1, \ldots, \mathbf{x}^n)$ as $\mu \to 0$. Consequently, using (2.47) and (2.42) gives

$$\Lambda^{(\mu)}(\mathbf{x} \mid \mathbf{x}^1, \ldots, \mathbf{x}^n) = \mu^{-n+s} \sum_{|I|=s} a_I \mid \det Y_I \mid^{-1} e^{\mu \mathbf{x}^I \bullet \mathbf{x}} \chi_{<I>_+}(\mathbf{x})$$

$$= \sum_{\ell=0}^{\infty} \frac{\mu^{\ell-n+s}}{\ell!} \left(\sum_{|I|=s} (\mathbf{x}^I \bullet \mathbf{x})^\ell a_I \mid \det Y_I \mid^{-1} \chi_{<I>_+}(\mathbf{x}) \right).$$

Since the left hand side of the above identity remains bounded for every $\mathbf{x}$ as μ tends to zero we conclude

$$\sum_{|I|=s} (\mathbf{x}^I \cdot \mathbf{x})^\ell a_I \ | \det Y_I |^{-1} \chi_{<I>_+}(\mathbf{x}) = 0, \ \ell = 0, \ldots, n-s-1 ,$$

and

$$T(\mathbf{x} \mid \mathbf{x}^1, \ldots, \mathbf{x}^n) = \frac{1}{(n-s)!} \sum_{|I|=s} a_I \ | \det Y_I |^{-1} (\mathbf{x}^I \cdot \mathbf{x})^{n-s} \chi_{<I>_+}(\mathbf{x}). \tag{2.49}$$

This formula reduces to (2.23) when $s = 1$.

We finish this section one more formula. In [7], we pointed out that

$$B(\,\cdot \mid X) = \nabla_X T(\,\cdot \mid X) \tag{2.50}$$

where ∇_X is a difference operator defined as

$$\nabla_X f(\,\cdot\,) = \prod_{y \in X} \nabla_y f(\,\cdot\,)$$
$$\nabla_y f(\,\cdot\,) = f(\,\cdot\,) - f(\,\cdot - y).$$

The proof of (2.50) follows directly from the defining equations (2.21)-(2.22) for the box spline and truncated power. Since it is obvious that

$$\nabla_X f(\,\cdot\,) = \sum_{V \subset X} (-1)^{|V|} f\left(\,\cdot\, - \sum_{x \in V} x\right)$$

we get

$$B(x \mid X) = \sum_{V \subset X} (-1)^{|V|} T\left(x - \sum_{x \in V} x \mid X\right).$$

In the univariate case we can use (2.23) and reduce the above identity to

$$B(t \mid x_1, \dots, x_n) = \frac{1}{(n-1)! x_1 \dots x_n} \sum_{V \subset \{x_1, \dots, x_n\}} (-1)^{|V|} \left(t - \sum_{x \in V} x\right)_+^{n-1}$$

when $x_1, \dots, x_n$ positive. Some remarks on the history of this formula are given in [9]. One should also note here that for $s = 1$, $X = \{1, \dots, 1\}$ the box spline is the same as the forward B-spline given by (2.4).

Much more is known about box splines. For instance, this function has surprising connections to the combinatorial theory of partitions.

3. Order Statistics.

The relationship between univariate B-splines and order statistics was recently pointed out in [12,13]. We start by supposing that $(t_1, \dots, t_n)$ is a sample from a uniformly distributed population. Let $t_1, \dots, t_n$ be rearranged in decreasing order and let the ordered values be $t_{(1)} > \cdots > t_{(n)}$. The corresponding new random variables $T_{(1)} \dots T_{(n)}$ are called the order statistics of the sample. Since $\mathbf{T} = (T_1 \dots T_n)$ is uniformly distributed over $(0, 1)^n$ and since for the order statistics all those samples are identified whose components agree up to a permutation, $(T_{(1)} \dots T_{(n)})$ is uniformly distributed over the simplex

$$\Sigma^n = \{u \in [0, 1]^n : u_1 \geq \cdots \geq u_n\}.$$

Now to find for given vectors $\mathbf{y}^1, \dots, \mathbf{y}^n \in \mathbb{R}^s$ the density $V(\mathbf{x} \mid \mathbf{y}^1, \dots, \mathbf{y}^n)$ of the random vector

$$\mathbf{X} = \mathbf{y}^1 T_{(1)} + \cdots + \mathbf{y}^n T_{(n)}$$

we note that the expectation of the random variable $f(\mathrm{X})$ is given by

$$\int_{\mathbb{R}^s} f(\mathbf{x})\, V(\mathbf{x} \mid \mathbf{y}^1, \ldots, \mathbf{y}^n)\, d\mathbf{x} = n! \int_{\Sigma^n} f(t_1 \mathbf{y}^n + \cdots + t_n \mathbf{y}^n) dt_1 \ldots dt_n. \tag{3.1}$$

Rewriting the right hand side as

$$n! \int_{\Delta_n} f(t_1 \mathbf{y}^1 + t_2(\mathbf{y}^1 + \mathbf{y}^2) + \cdots + t_n(\mathbf{y}^1 + \cdots + \mathbf{y}^n))\, dt_1 \ldots dt_n$$

and recalling definition (2.18) gives the density of $\mathbf{X}$ as a multivariate B-spline

$$V(\mathbf{x} \mid \mathbf{y}^1, \ldots, \mathbf{y}^n) = M(\mathbf{x} \mid 0, \mathbf{y}^1, \mathbf{y}^1 + \mathbf{y}^2, \ldots, \mathbf{y}^1 + \cdots + \mathbf{y}^n). \tag{3.2}$$

Remark: The distribution of the order statistics $(T_{(1)} \ldots T_{(n)})$ is usually referred to as the ordered Dirichlet distribution $D(1, \ldots, 1; 1)$ where in general $D(\alpha_1, \ldots, \alpha_n; \alpha_{n+1})$ corresponds to the density

$$\chi_{\Sigma^n}(\tau) \frac{\Gamma(\alpha_1 + \cdots + \alpha_{n+1})}{\Gamma(\alpha_1) \ldots \Gamma(\alpha_{n+1})} \tau_1^{\alpha_1 - 1} (\tau_2 - \tau_1)^{\alpha_2 - 1} \ldots (\tau_n - \tau_{n-1})^{\alpha_n - 1} (1 - \tau_n)^{\alpha_{n+1} - 1},$$

Cf. Wilks [27].

We can also interpret $V(\,\bullet \mid \mathbf{y}^1, \ldots, \mathbf{y}^n)$ in (3.2) as the conditional density of the random vector

$$\mathbf{X} = T_1 \mathbf{y}^1 + \cdots + T_n \mathbf{y}^n$$

subject to the condition $T_1 \geq \cdots \geq T_n$, Wilks [27]. More generally, denoting by $\mathcal{P}_n$ the set of all permutations of $\{1, \ldots, n\}$ we may define for any $\pi \in \mathcal{P}_n$, $V_\pi(\mathbf{x} \mid \mathbf{y}^1, \ldots, \mathbf{y}^n)$ to be the conditional density of the random vector

$$\mathbf{X} = T_1\mathbf{y}^1 + \cdots + T_n\mathbf{y}^n$$

subject to the condition $T_{\pi(1)} \geq \cdots \geq T_{\pi(n)}$ giving as above the formula

$$V_\pi(\mathbf{x} \mid \mathbf{y}^1, \ldots, \mathbf{y}^n) = M(\mathbf{x} \mid 0, \mathbf{y}^{\pi(1)}, \ldots, \mathbf{y}^{\pi(1)} + \cdots + \mathbf{y}^{\pi(n)}). \tag{3.3}$$

Now the (unconditional) density of the linear combination $\mathbf{X} = T_1\mathbf{y}^1 + \cdots + T_n\mathbf{y}^n$ is the box spline $B(\mathbf{x} \mid \mathbf{y}^1, \ldots, \mathbf{y}^n)$. On the other hand, since the density of $T_1\mathbf{y}^1 + \cdots + T_n\mathbf{y}^n$ can also be obtained by summing over all conditional densities corresponding to any permutation in $\mathcal{P}_n$ we obtain a formula for the box spline in terms of the multivariate B-spline

$$B(\mathbf{x} \mid \mathbf{y}^1, \ldots, \mathbf{y}^n) = \frac{1}{n!} \sum_{\pi \in \mathcal{P}_n} M(\mathbf{x} \mid 0, \mathbf{y}^{\pi(1)}, \ldots, \mathbf{y}^{\pi(1)} + \cdots + \mathbf{y}^{\pi(n)}). \tag{3.4}$$

4. Urn Models.

Our final example of a statistical interpretation of B-splines comes from recent work of Goldman [10]. His motivation comes from certain mathematical questions connected with the computer generation of curves and surfaces. An extremely useful and powerful method which provides a solution to this problem is to represent a curve $\mathbf{x}(t)$, $0 \leq t \leq 1$, as

$$\mathbf{x}(t) = \sum_{j=0}^{n} \mathbf{V}^j P_{j,n}(t). \tag{4.1}$$

The vectors $\mathbf{V}^0, \mathbf{V}^1, \ldots, \mathbf{V}^n$ are referred to as **control points** which are typically supplied by the designer and $P_{j,n}(t)$ are real valued functions called **blending functions.** The polygonal curve obtained by successively joining $\mathbf{V}^0, \ldots, \mathbf{V}^n$ by straight lines is called the **control polygon** which represents a "primitive" version of the desired curve. The blending functions are therefore a device to provide a "smooth" rendering of the control polygon. Goldman [10] argues that simple and natural requirements on this representation of the control polygon leads to the conditions

$$P_{j,n}(t) \geq 0, \quad \sum_{j=0}^{n} P_{j,n}(t) = 1, \quad 0 \leq t \leq 1. \tag{4.2}$$

We recognize these equations as the characteristic properties of a discrete probability distribution. Thus Goldman suggests that probability and computer aided design (CAD) are intimately related and proposes to generate blending functions from urn models. We now describe a specific instance of this procedure which leads to B-splines.

The urn model we consider includes the Friedman urn model studied in [10] and is described in Johnson and Kotz (1977), [15]. Following their description we begin with an urn containing w white balls and b black balls. A ball is drawn at random from the urn and its color is determined. It is then returned and a constant number of new balls is added to the urn according to the rule

		Color of Chosen Ball	
		White	Black
Number of balls added to urn	White	ω_w	ω_b
	Black	β_w	β_b

(4.3)

The Friedman urn model corresponds to the case that $\beta_b = \omega_w$, $\beta_w = \omega_b$. We let $t = \dfrac{w}{w+b}$ and introduce

$$P_{j,n}(t) = \text{probability that } j \text{ white balls are drawn after } n \text{ trials.} \tag{4.4}$$

The main tool we use to study these functions is the recurrence relation

$$P_{j,n+1}(t) = \frac{1 - t + j\tilde{\beta}_w + (n-j)\tilde{\beta}_b}{1 + j\tilde{\beta}_w + (n-j)\tilde{\beta}_b + j\tilde{\omega}_w + (n-j)\tilde{\omega}_b} P_{j,n}(t)$$
$$+ \frac{t + (j-1)\tilde{\omega}_w + (n-j+1)\tilde{\omega}_b}{1 + (j-1)\tilde{\omega}_w + (n-j+1)\tilde{\omega}_b + (j-1)\tilde{\beta}_w + (n-j+1)\tilde{\beta}_b} P_{j-1,n}(t) \tag{4.5}$$

which holds for $j = 0, 1, \ldots, n+1$. Here the tilde variables are obtained by dividing their un-tilde counterparts by $b + w$.

For the proof of this formula we observe that j white balls can be obtained in $n + 1$ trials in only two ways. At the n-th trial we have already selected j white balls and the color of the next ball is black. Otherwise, at the n-th trial we only have selected $j - 1$ white balls and the color of the next selection is white. Now, the probability of choosing a black ball in the $n + 1^{\text{st}}$ trial in the first case above is

$$\frac{b + j\beta_w + (n-j)\beta_b}{b + w + j\beta_w + (n-j)\beta_b + j\omega_w + (n-j)\omega_b} \tag{4.6}$$

while the probability of choosing a white ball in the $n + 1^{\text{st}}$ trial in the second case is

$$\frac{w + (j-1)\omega_w + (n-j+1)\omega_b}{b + w + (j-1)\beta_w + (n-j+1)\beta_b + (j-1)\omega_w + (n-j+1)\omega_b}. \tag{4.7}$$

Consequently, the recurrence (4.5) follows. Observe that $P_{0,1}(t) = 1 - t$ and $P_{1,1}(t) = t$ which implies inductively from (4.5) that each $P_{j,n}$ is a polynomial of at most degree n. This suggests the following B-spline ansatz for the identification of probabilities

$$P_{j,n}(t) = N(t \mid t_{-n+j}, \ldots, t_{j+1}), \quad 0 \le t \le 1, j = 0, 1, \ldots, n, \tag{4.8}$$

for some knots $t_{-n} \le \cdots < t_{n+1}$ where $t_0 = 0$, $t_1 = 1$ and $N(t \mid t_{-n+j}, \ldots, t_{j+1}) = (t_{j+1} - t_{-n+j}) M(t \mid t_{-n+j}, \ldots, t_{j+1})$ is the B-spline normalized to satisfy (4.2), Cf. [2,24]. When equation (4.8) holds we can view $P_{j,n}$ as the "pieces" of the $j + 1^{st}$ normalized B-spline on $[0,1]$ for the given knot sequence. The reason that the normalized B-splines in (4.8) might be urn model probabilities is that they satisfy the formula

$$P_{j,n+1}(t) = \frac{t - t_{-n+j-1}}{t_j - t_{-n+j-1}} P_{j-1,n}(t) + \frac{t_{j+1} - t}{t_{j+1} - t_{-n+j}} P_{j,n}(t), \quad j = 0, 1, \dots, n+1, \tag{4.9}$$

Cf. [2,24]. This leads us to the following fact.

The urn model above has probabilities corresponding to normalized B-splines if and only if

$$\beta_b = \omega_w = 0 \text{ and } \omega_b = \beta_w. \tag{4.10}$$

Moreover, in this case the knots are given by

$$t_j = \begin{cases} 1 + (j-1)\tilde{\beta}_w, & j = 1, \dots, n+1, \\ j\tilde{\beta}_w, & j = -n, \dots, 0. \end{cases} \tag{4.11}$$

For the proof of this result we observe that the recurrence relations (4.8) and (4.9) are identical if and only if

$$t_j - t_{-n+j-1} = 1 + (j-1)\tilde{\omega}_w + (n-j+1)\tilde{\omega}_b + (j-1)\tilde{\beta}_w + (n-j+1)\tilde{\beta}_b, \quad j = 1, \dots, n+1, \tag{4.12}$$

$$t_{j+1} - t_{-n+j} = 1 + j\tilde{\omega}_w + (n-j)\tilde{\omega}_b + j\tilde{\beta}_w + (n-j)\tilde{\beta}_b, \quad j = 0, 1, \dots, n, \tag{4.13}$$

and

$$t_{j+1} = 1 + j\tilde{\beta}_w + (n-j)\tilde{\beta}_b, \quad j = 0, 1, \dots, n. \tag{4.14}$$

The first two relations lead to the requirement that

$$\beta_b + \omega_b = \beta_w + \omega_w, \tag{4.15}$$

that is, the total number of balls added is the same whether a black or a white ball is chosen.

We get from (4.14) that

$$t_j = 1 + (j-1)(\tilde{\beta}_w - \tilde{\beta}_b) + n\tilde{\beta}_b, \quad j = 1, \dots, n+1.$$

Thus, in particular $t_1 = 1 + n\tilde{\beta}_b$ and $t_{n+1} = 1 + n\tilde{\beta}_w$ which means $\tilde{\beta}_b = 0$ because $t_1 = 1$. However, we also require $t_0 = 0$ which gives from (4.13) when $j = n$ that $\omega_w = 0$. We can now solve for the remaining negative knots and thereby complete the proof.

This result identifies the probabilities $P_{j,n}$ when (4.10) holds, as we may use the divided difference formula (2.7) for the B-spline. Of course, in the case that

$$\omega_w = \omega_b = \beta_w = \beta_b = 0,$$

that is, no balls are added, we get the binomial distribution

$$P_{j,n}(t) = \binom{n}{j} t^j (1-t)^{n-j}, \quad j = 0, 1, \dots, n,$$

in agreement with our formula (2.36) ($\alpha_0 = \cdots = \alpha_n = s = 1$). These blending functions were used by P. Bézier and are the basis for the general point of view for curve generation given by (4.1). They have been found to be very useful in CAD.

Finally, we mention that when $\beta_b = \omega_w$ and $\beta_w = \omega_b = 0$ then we get the Pólya-Eggenberger urn model [15]. In this case, an explicit formula for the probabilities is known

$$P_{j,n}(t) = \frac{\binom{n}{j} \prod_{\ell=0}^{j-1} (t + (\ell - 1)\beta_b) \prod_{\ell=0}^{n-j-1} (1 - t + (n - \ell - 1)\beta_b)}{\prod_{\ell=0}^{n-1} (1 + j\beta_b)} .$$

These polynomials are closely tied to the B-spline model described above by the basic identity

$$(y - t)^n = \sum_{i=0}^{n} (y - t_{-n+j+1}) \cdots (y - t_{j-1}) N(t \mid t_{-n+j}, \ldots, t_j),$$

where the knots are given by (4.11), see [2,24]. It would be interesting to have a probabilistic explanation for this formula.

REFERENCE:

1. R.L. Anderson (1942). Distribution of serial correlation coefficient, **Ann. Math. Statist. 13,** 1-13.

2. C. de Boor (1978). **A Practical Guide to Splines,** Springer-Verlag, Berlin-Heidelberg.

3. C. de Boor (1976). Splines as linear combinations of B-splines in: **Approximation Theory II,** eds. G.G. Lorentz, C.K. Chui and L.L. Schumaker, Academic Press, New York, 1-47.

4. H.B. Curry, I.J. Schoenberg, (1947). On spline distributions and their limits: the Pólya distribution functions, abstract 380T, **Bull. AMS 53**, 1114.

5. H.B. Curry, I.J. Schoenberg, (1966). On Pólya frequency functions. IV. The fundamental spline functions and their limits, **J. d'Analyse Math. 17,** 71-107.

6. W. Dahmen, C.A. Micchelli (1981). On the limits of multivariate B-splines, **J. d'Analyse Math.** 39, 256-278.

7. W. Dahmen, C.A. Micchelli (1983). Recent progress in multivariate splines, in **Approximation Theory** IV, eds. C.K. Chui, L.L. Schumaker, J.D. Ward, Academic Press, New York, 17-121.

8. A.P. Dempster, R.M. Kleyle (1968). Distributions determined by cutting a simplex with hyperplanes, **Ann. Math. Statist., 39,** , 1473-8.

9. P. Diaconis, B. Efron (1986). Probabilistic geometric theorems arising from the analysis of contingency tables, preprint.

10. R. Goldman (1985). Pólya urn model and computer aided geometric design, **J. Alg. Disc. Meth. 6,** 1-28.

11. C.L. Hu, S.L. Lee (1983). Multivariate B-splines in the joint distributions of the circular serial correlation coefficients, preprint.

12. Z.G. Ignatov, V.K. Kaishev (1985). B-splines and linear combinations of uniform order statistics, MRC technical summary report #2817, University of Wisconsin, Madison, Wisconsin.

13. Z.G. Ignatov, V.K. Kaishev (1985). A probabilistic interpretation of multivariate B-splines and some applications, to appear in Annals of Probability.

14. N.L. Johnson, S. Kotz (1972). **Distributions in Statistics: Continuous Multivariate Distributions** John Wiley & Sons, New York.

15. N.L. Johnson, S. Kotz (1977). **Urn Models and Their Applications,** John Wiley & Sons, New York.

16. S. Karlin, C.A. Micchelli, Y. Rinott (1984). Multivariate splines: a probabilistic perspective, to appear **J. Multivariate Analysis.**

17. T. Koopmans (1942). Serial correlation and quadratic forms in normal variables, **Ann. Math. Statist. 13,** 14-33.

18. P.S. Laplace (1820). Théorie Analytique des Probabilities. (3rd Edition) Courcier, Paris.

19. J. von Neumann (1941). Distribution of the ratio of the mean square successive differences to the variance, **Ann. Math. Statist. 12,** 367-395.

20. J. von Neumann (1942). A further remark concerning the distribution of the ratio of the mean square successive difference to the variance, **Ann. Math. Statist. 13,** 86-88.

21. M.H. Quenouille (1949). The joint distribution of serial correlation coefficients, **Ann. Math. Statist. 20,** 561-571.

22. I.J. Schoenberg (1946). Contributions to the problem of approximation of equidistant data by analytic functions, part A, **Quart. Appl. Math. 4,**45-99; part B, **ibid 4,** 112-141.

23. I.J. Schoenberg (5/31/65). A letter to P.J. Davis, unpublished.

24. L.L. Schumaker (1981). **Spline Functions,** John Wiley & Sons, New York.

25. G.S. Watson (1956). On the joint distribution of circular serial correlation coefficients, **Biometrika 4,** 161-168.

26. E.J. Wegman, I.W. Wright (1983). Splines in statistics, **J. Amer. Statist. Assoc. 78,** 351-365.

27. S.S. Wilks (1962). **Mathematical Statistics**, John Wiley & Sons, New York.

Universität Bielefeld
Fakultät für Mathematik
4800 Bielefeld 1
West Germany

Department of Mathematical Sciences
IBM Thomas J. Watson Research Center
P.O. Box 218
Yorktown Heights, NY 10598

Contemporary Mathematics
Volume **59**, 1986

ESTIMATION OF A TRANSFER FUNCTION IN A NONGAUSSIAN CONTEXT

K. S. Lii and M. Rosenblatt

ABSTRACT. Aspects of a deconvolution problem and estimation of an associated transfer function are considered in the context of a nonGaussian linear process.

INTRODUCTION. In a number of geophysical problems the following model has been proposed. Consider a sequence of independent, identically distributed random variables η_t, $t = \ldots, -1, 0, 1, \ldots$, with $E\eta_t \equiv 0$, $E\eta_t^2 \equiv 1$, that is nonGaussian. Let $\{\alpha_j\}$ be a sequence of real constants with

$$\sum \alpha_j^2 < \infty \quad .$$

The sequence $\{\eta_t\}$ is convolved with the sequence $\{\alpha_j\}$ to produce an output sequence

$$x_t = \sum_j \alpha_j \, \eta_{t-j} \quad .$$

This is a context in which the model has been proposed. Consider a layered earth. An explosion is set up on the surface with the $\{\alpha_j\}$ sequence (the wavelet) representing a pulse set up by the explosion. The η_t's are the reflectivity coefficients at the interface of the different layers. The x_t sequence is recorded at the surface by a geophone and from it one wishes to infer as much as possible (often without prior knowledge) about the wavelet sequence $\{\alpha_j\}$ and the reflectivity sequence $\{\eta_t\}$.

The spectral density of the process $\{x_t\}$ is

$$f(\lambda) = \frac{1}{2\pi} |\alpha(e^{-i\lambda})|^2 \quad ,$$

with

$$\alpha(e^{-i\lambda}) = \sum_j \alpha_j e^{-ij\lambda} \quad ,$$

the transfer function from $\{\eta_t\}$ to $\{x_t\}$. In the Gaussian case knowledge of the spectral density and knowledge of the full probability structure are equivalent.

1980 Mathematics Subject Classification (1985 Revision). 62M15, 73D50, 73N99.
This research was supported in part by Office of Naval Research contract N00014-81-K-003 and National Science Foundation grant DMS-83-12106.

However, from the knowledge of $f(\lambda)$ one can only determine the modulus of α, $|\alpha(e^{-i\lambda})|$. For this reason, in the Gaussian case, to determine $\alpha(e^{-i\lambda})$ one usually makes an ad hoc assumption called the minimum phase assumption. This essentially assumes that the real and imaginary parts of $\alpha(e^{-i\lambda})$ are conjugate functions of each other. If

$$\alpha(e^{-i\lambda}) = \frac{a(e^{-i\lambda})}{b(e^{-i\lambda})}$$

where $a(z)$, $b(z)$ are polynomials in z, this amounts to the assumption that the roots of $a(z)$ and $b(z)$ are outside the unit disc in the complex plane. In the nonGaussian case, under appropriate conditions, one can determine almost all information about $\alpha(e^{-i\lambda})$ asymptotically as $n \to \infty$, given that one has a sample $x_1, \ldots, x_n$ of the process $\{x_t\}$. A number of papers discuss the analysis of the nonGaussian model [1,3-9]. The paper of Godfrey and Rocca [2] proposes another related nonGaussian model.

GENERAL DISCUSSION. Conditions of the following character have often been made (see [5]) to obtain the desired results. A first assumption is that a kth order moment $(k > 2)$ for $\{\eta_t\}$ exists with corresponding k^{th} cumulant nonzero. One also requires that

$$\sum |j| |\alpha_j| < \infty \tag{1}$$

with

$$\alpha(e^{-i\lambda}) \not\equiv 0 \tag{2}$$

for all λ. However, we should remark that this assumption on the nonzero character of $\alpha(e^{-i\lambda})$ can be weakened somewhat. The higher moment assumption is used because it is this in part that allows one to determine most of the information about the phase of $\alpha(e^{-i\lambda})$.

Under the assumption that the kth order cumulant γ_k of $\{\eta_t\}$ is nonzero, the kth order cumulant

$$\begin{aligned} &\operatorname{cum}(x_{t_0}, x_{t_1}, \ldots, x_{t_{k-1}}) \\ &= \gamma_k \sum_s \alpha_s \alpha_{s+t_1-t_0} \cdots \alpha_{s+t_{k-1}-t_0} \end{aligned}$$

and the kth order cumulant spectral density of the process x_t is

$$\begin{aligned} b_k(\lambda_1, \ldots, \lambda_{k-1}) &= \frac{\gamma_k}{(2\pi)^{k-1}} \sum_{j_1, \ldots, j_{k-1}} \operatorname{cum}(x_0, x_{j_1}, \ldots, x_{j_{k-1}}) \exp\left[-\sum_{b=1}^{k-1} i j_b \lambda_b\right] \\ &= \frac{\gamma_k}{(2\pi)^{k-1}} \alpha(e^{-i\lambda_1}) \cdots \alpha(e^{-i\lambda_{k-1}}) \, \alpha\left(e^{i(\lambda_1 + \cdots + \lambda_{k-1})}\right). \end{aligned}$$

We introduce the normalization

$$h(\lambda) = \arg\left\{\alpha(e^{-i\lambda}) \frac{\alpha(1)}{|\alpha(1)|}\right\}$$

assuming that $\alpha(1) \neq 0$. If we assume $\alpha(1) > 0$, so that $h(0) = 0$, then

$$h(\lambda_1) + \cdots + h(\lambda_{k-1}) - h(\lambda_1 + \cdots + \lambda_{k-1})$$

$$= \arg\left\{\left[\frac{\alpha(1)}{|\alpha(1)|}\right]^k \gamma_k^{-1}\, b_k(\lambda_1, \ldots, \lambda_{k-1})\right\} \tag{3}$$

and note that $h(-\lambda) = -h(\lambda)$ since the coefficients α_j's are real. A similar analysis can be given in case $\alpha(1) < 0$. Notice that

$$h'(0) - h'(\lambda) = \lim_{\Delta \to 0} \frac{1}{(k-2)\Delta}\left\{h(\lambda) + (k-2)h(\Delta) - h(\lambda + (k-2)\Delta)\right\} \tag{4}$$

and that

$$h(\lambda) = \int_0^{\lambda}\left\{h'(u) - h'(0)\right\}du + c\lambda \tag{5}$$

with $c = h'(0)$. By making use of (3), (4), and (5), and estimates of the kth order cumulant spectrum $b_k(\lambda_1, \ldots, \lambda_{k-1})$ one can estimate the phase of $\alpha(e^{-i\lambda})$ up to a linear factor. Using this and an estimate of the spectral density one can estimate $\alpha(e^{-i\lambda})$ up to an undetermined sign and factor $e^{in\lambda}$ with n an integer.

It is of some interest to see to what extent the conditions (1) and (2) can be relaxed. In the paper of Lii and Rosenblatt 1986 [6], a partial relaxation of the condition (2) on the zeros is noted.

REFERENCES

1. D. Donoho, "On minimum entropy deconvolution," Applied Time Series Analysis II (ed. D. F. Findley) (19881), 565-608.

2. R. Godfrey and F. Rocca, "Zero memory nonlinear deconvolution," Geophysical Prospecting **29** (1981), 189-228.

3. K. S. Lii and M. Rosenblatt, "A fourth-order deconvolution technique for nonGaussian linear processes," Multivariate Analysis VI (ed. P. R. Krishnaiah), Elsevier, 1985, pp. 395-410.

4. K. S. Lii and M. Rosenblatt, "NonGaussian linear processes, phase and deconvolution," Statistical Signal Processing, (eds. E. J. Wegman and J. G. Smith), M. Dekker 1984, pp. 51-58.

5. K. S. Lii and M. Rosenblatt, "Deconvolution and estimation of transfer function phase and coefficients for nonGaussian linear processes," Ann. Statist. **10** (1982), 1195-1208.

6. K. S. Lii and M. Rosenblatt, "Deconvolution of nonGaussian linear processes with vanishing spectral values," Proc. Nat. Acad. Sci. USA, 1986.

7. T. Matsuoka and T. J. Ulrych, "Phase estimation using the bispectrum," Proc. of IEEE **72** (1984), 1403-1411.

8. M. Rosenblatt, "Linear processes and bispectra," J. Appl. Prob., **17** (1980), 265-270.

Department of Mathematics
University of California, Riverside
Riverside, California 92521

Department of Mathematics
University of California, San Diego
La Jolla, California 92093

Contemporary Mathematics
Volume **59**, 1986

Evaluating the Performance of an Inversion Algorithm

Finbarr O'Sullivan[1]

ABSTRACT. An approach to understanding the small sample spatial performance characterists of a given inversion algorithm is described. The method involves a generalization of the Backus-Gilbert formulism and yields useful assessments of bias and variability. An extension to non-linear inversion algorithms is also indicated. The method is illustrated in the context of the History Matching problem of reservoir engineering.

1. Introduction. A broad class of inverse problems are formulated in connection with the estimation of (functional) parameters in partial differential equations, see Anger[1], Lions[8], McLaughlin[9], and especially Payne[13]. The History Matching problem of reservoir engineering is concerned with making inferences about certain reservoir parameters based on data accumulated in time at distributed well sites. Chapter 1 of Peaceman[14] gives an overview of the basic equations used to model reservoirs. Typical hydrocarbon reservoirs are multi-phase, involving mixtures of oil, gas and water. In a single phase system, such as an aquifer, the flow of fluid is modeled by the diffusion equation

$$\frac{\partial u(\mathbf{x},t)}{\partial t} - \frac{\partial}{\partial \mathbf{x}}\left\{ a(\mathbf{x})u(\mathbf{x},t)\right\} = q(\mathbf{x},t) \qquad \mathbf{x} \text{ in } \Omega \text{ and } t \text{ in } [0,T] ,$$

subject to prescribed initial and boundary conditions. u is pressure, q accounts for the withdrawal or injection of fluid into the region Ω, and a is the transmissivity or

1980 Mathematics Subject Classification (1985 Revision). 65R20, 65D10, 62G05.

Key-words and phrases. averaging kernel, history matching, ill-posed inverse problems, mean square error.

[1] This work was supported by the National Science Foundation under Grant Number MCS 84-03239.

permeability which determines the ease with which fluid flows through the reservoir. The initial condition is $u(\mathbf{x},0) = u_o(\mathbf{x})$ and a usual boundary condition is that there is no fluid flow across the boundary of the region, i.e. $\frac{\partial u}{\partial \omega} = 0$, where ω represents the direction normal to the boundary.

In the standard History Matching problem one tries to use scattered well data on $u(\mathbf{x}_i, t_j)$ and $q(\mathbf{x}_i, t_j)$, for $i=1,2, \cdots, n$ and $j=1,2, \cdots, T$, to infer the (functional) parameter, a. The term History Matching derives from the notion that a good specification for the reservoir parameters should result in predictions which closely match the observed pressure history data. The problem is ill-posed, in the sense of Hadamard[6], and some form of regularization is needed in order to produce meaningful solutions. A variety of *Inversion Methods* or History Matching algorithms have been proposed in both the Hydrology (Cooley[4] and Neuman and Yakowitz[10]) and Petroleum Engineering (Kravaris and Seinfeld[7]) literature. The practical performance of these algorithms is not well understood. At the same time, the ability to assess spatial performance characteristics for a given algorithm and well-site configuration is very important. This is especially true for off-shore petroleum reservoirs where the costs associated with the construction of wells are huge.

The goal of my talk was to highlight a simple calculus which provides some valuable insight into the performance of history matching type algorithms. The basic method is defined and illustrated below. An important feature of the method is that it is non-asymptotic. In many inverse problems n is *very small,* in the History Matching problem, for example, the number of well sites may only be about 10.

2. Basic Calculus. Let θ denote the functional target parameter (in the History Matching problem θ could be the reservoir permeability). Assume that θ lies in an appropriate Hilbert space Θ with inner product $<\cdot,\cdot>$, and that, for any point t in the domain of θ, evaluation at t is a continuous linear functional. An ***inversion algorithm,*** S, is a mapping from the data space into Θ. Formally,

$$\hat{\theta} = S\{\ data\ \}\ .$$

The goal is to try and understand the behavior of the inversion algorithm at points t of interest. This discussion focuses on the mean square error (MSE) performance,

$$\text{MSE}(t) = \text{E}\ [\ \theta(t) - \hat{\theta}(t)]^2\ .$$

2.1. Linear Case. By a linear problem I mean a situation in which the the measurement system provides information on linear functionals, perhaps integrals, of the target parameter and the inversion algorithm, S, is linear. Thus

$$z_i = X_i \theta + \epsilon_i \quad , \quad i = 1,2, \cdots ,n \ ,$$

with the X_i's continuous linear functionals and the ϵ_i's independent mean zero measurement errors with common variance σ^2. Since the inversion algorithm is a linear mapping from the data space, R^n, into the parameter space, Θ,

$$\hat{\theta} = S \underline{z} = \sum_{i=1}^{n} S^{(i)} z_i \quad .$$

$S^{(i)}$ are the impulse response functions for the algorithm. Θ is a Hilbert space and the functionals X_i are continuous, so $\hat{\theta}(t)$ has the representation

$$\hat{\theta}(t) = <A(t),\theta> + \sum_{i=1}^{n} S^{(i)}(t)\epsilon_i \tag{2.1}$$

where

$$A(t) = \sum_{i=1}^{n} S^{(i)}(t)\xi_i \quad ,$$

and ξ_i is the Riesz representer of the functional X_i in Θ. In particular, if the inner product on Θ is L_2, then

$$\mathrm{E}\, \hat{\theta}(t) = \int A(t,s)\, \theta(s)\, ds \quad .$$

Here the function $A(t,\cdot)$ describes the degree of local averaging inherent in the retrieved estimate at the point t, and coincides with what Geophysicists call the Backus-Gilbert[2] averaging kernel. For other inner products, $A(t)$ is a natural generalization of the Backus-Gilbert formulation.

The representation (2.1), allows one to separate out the bias and variability components of the estimate. Because the evaluation functional is assumed continuous at t, the Cauchy-Schwartz inequality implies that the maximum bias, b_M, incurred over all functions in a ball of radius μ is proportional to the norm of the difference between the representer of evaluation, η_t, and the function $A(t)$.

$$b_M^2(t) = \sup_{||\theta|| \le \mu} [\theta(t) - \mathrm{E}\, \hat{\theta}(t)]^2 = ||\eta_t - A(t)||^2 \mu^2 \quad .$$

Meanwhile the variability at t is

$$var(t) = \sigma^2 \sum_{i=1}^{n} S^{(i)}(t)^2 .$$

Combining the bias and variability provides a description of the worst case mean square error at t. Average case performance with respect to a prior distribution on the set Θ could be computed in a similar fashion.

Example 1. An Integral Equation in Stereology

Suppose the data functionals are integrals.

$$X_i\,\theta = \int_0^1 K(x_i, t)\,\theta(t)\,dt \quad , \quad i = 1,2,\cdots,n ,$$

where the kernels, $K(x_i, \cdot)$, are given by

$$K(x_i, t) = \begin{cases} 0 & 0 \le t < x_i \\ \sqrt{t^2 - x_i^2} \quad , & x_i \le t < x_{i+1} \\ \sqrt{t^2 - x_i^2} - \sqrt{t^2 - x_{i+1}^2} & x_{i+1} \le t < 1 \end{cases}$$

Such kernels arise in connection with a discretized integral equation in stereology, see Nychka et al.[11]. Let estimators of θ be obtained by Tikhonov's Method of Regularization (MOR), $\hat{\theta}$ is the minimizer in $W_2^2\,[0,1]$ of

$$\sum_{i=1}^{n} [z_i - \int_0^1 K(x_i, t)\,\theta(t)\,dt\,]^2 + \lambda \int_0^1 [\ddot{\theta}(t)]^2\,dt \quad , \quad \lambda > 0. \tag{2.2}$$

The functional is quadratic, and it is easily shown that $\hat{\theta}$ is linear in the data. n equal to 50 equi-spaced x_i's are considered. For a fixed value of λ, the L_2 averaging kernel at $t = .4$ is given in Figure 2.1. The corresponding maximum absolute bias, with respect to the Sobolev $W_2^2\,[0,1]$- norm, and standard error (square root of the variance) are given in Figure 2.2. The vertical scale in the plots are not included but these are proportional to the true bias and standard error characteristics of the inversion algorithm. Note that the inversion characteristics degrade for small values of t - this is consistent with the shape of the kernels. A shrewd experimental design might put more effort into sampling near the left end-point. The ripples in the bias plot is due to the finite sampling. In the above evaluation λ is fixed, optimally corresponding, in the method of regularization procedure, to an assumed noise-to-signal ratio[12]. Thus at the stage when one is evaluating the performance of the inversion algorithm, the averaging kernel calculus allows one to obtain performance characteristics for a range of possible noise-to-signal

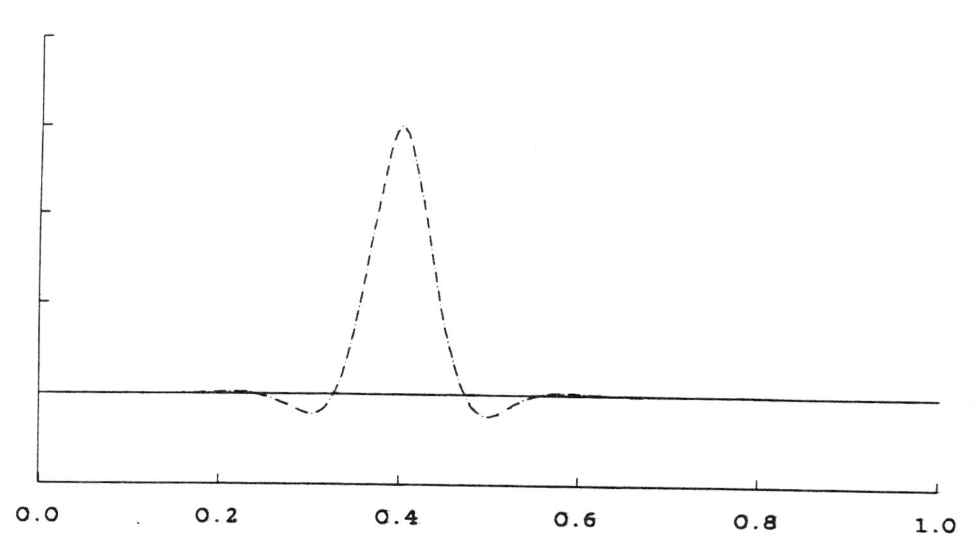

Figure 2.1 : A MOR L_2 Averaging Kernel at t =.4 for the Integral Equation example.

ratios. On real data λ must be empirically selected, typically to balance the bias and variability of the estimate. Methods such as cross-validation and unbiased risk have been developed for this purpose. There are several contributions in this volume devoted to this important topic.

2.2. Non-Linear Case. Non-linear inversion algorithms arise in many ways. For instance if measurements are made on non-linear functionals of θ, then any reasonable inversion algorithm must be non-linear. This is the case in History Matching: the data are of the form

$$z_i = \eta(x_i, \theta) + \epsilon_i \qquad \text{for} \quad i = 1, 2, \cdots, n \ , \tag{2.3}$$

where $\eta(x_i, \cdot)$'s are non-linear functionals of θ (the underlying permeability) and the ϵ_i's are measurement errors. Naturally, non-linearities complicate the analysis of the inversion algorithm. Although, one might think that computer simulation is the only available avenue, designing an adequate simulation is not an easy task. The variables are θ and the error distribution, and defining a test bed of θ values is particularly difficult. On the other hand, to apply the averaging kernel calculus it is necessary to linearize the inversion algorithm.

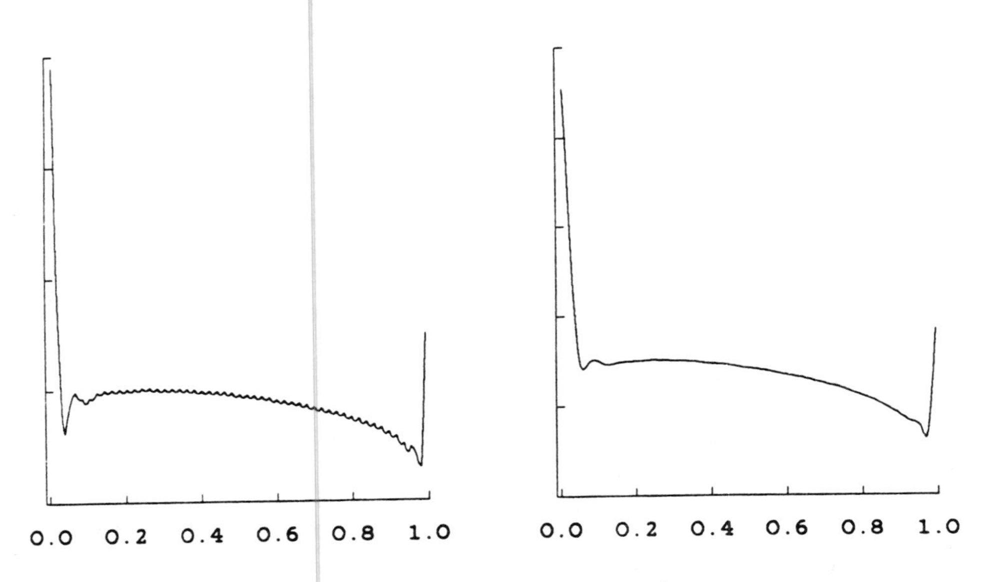

Figure 2.2 : Maximum Bias (left) and Standard Error (right) for the MOR applied to the Integral Equation example.

The basic steps in the linearized analysis can be illustrated in the context of a Method of Regularization algorithm. Here $\hat{\theta}$ is the minimizer over Θ of a functional of the form

$$\sum_{i=1}^{n} [z_i - \eta(x_i, \theta)]^2 + \lambda <\theta, W\theta> \quad , \quad \lambda > 0, \tag{2.4}$$

where W is some positive semi-definite linear operator. Linearizing the functionals about an element θ_0, gives

$$\eta(x_i, \theta) \approx \eta(x_i, \theta_0) + \nabla_\theta \, \eta(x_i, \theta_0) \, (\theta - \theta_0) \ .$$

Substituting into the regularization functional and minimizing with respect to θ, leads to a linear approximation to $\hat{\theta}$ of the form

$$\hat{\theta} \approx S_{\theta_0} \underline{z}^* \quad ,$$

where $z_i^* = z_i - \eta(x_i, \theta_0) + \nabla_\theta \, \eta(x_i, \theta_0) \, \theta_0$. The bias and variability properties of the operator S_{θ_0} can be analyzed by the averaging kernel calculus.

Example 2. A One-dimensional History Matching Problem

A simplified one-dimensional version of the history matching problem is considered by Kravaris and Seinfeld[7]. Let $\Omega = [0,1]$, assume there is no injection or withdrawal of fluid and that there is no flow across the boundary. The initial condition is

$$u_0(x) = 10 + 270x^2 - 180x^3 \quad .$$

There are 10 measurement sites at $x_i = (5i-3)/49$, $i=1,2, \cdots 10$, and at each site measurements on u are made for $t_j = 0.0(.007)0.5$. Since u is the solution to a diffusion equation involving the unknown permeability, θ, the measurements relate to non-linear functionals of θ.

Consider a method of regularization as in (2.4), with $<\theta, W\theta> = \int_0^1 [\ddot{\theta}(t)]^2 dt$. Linearizing the observed functionals about the permeability profile

$$\theta_0(x) = 0.5 + x - 5x^4 + 6x^5 - 2x^6 \quad ,$$

results in approximations to the bias and variability properties of the inversion algorithm. (The computational aspects of this program are not trivial, see O'Sullivan[12] for details). The results are displayed in Figure 2.3. For the given measurement site allocation, one can see that the quality of information retrieved by the algorithm, about the permeability function, degrades in a symmetric fashion as one moves to the edge of the region. The pronounced oscillations in the bias plot are due to the finite sampling.

It seems plausible that a linearized analysis ought to give good information on the bias and variability characteristics of the inversion algorithm whenever the *magnitude* of the non-linearities in the functionals $\eta(x_i, \cdot)$ are of small order. However, even for parametric non-linear regression models, there is a some difficulty in defining satisfactory measures of non-linearity for this purpose, see Cook and Witmer[3] for example. In certain circumstances, unfortunately not including (2.3), the theory in Cox and O'Sullivan[5] provides some asymptotic justification for linearization. It will be interesting to see if this work can be extended to the present situation.

3. Discussion. The averaging kernel calculus is a powerful tool yielding great insight into the nature of inversion algorithms. The method allows one compute the spatial performance of the algorithm and this information has many potential uses. Most importantly, information of this sort can provide guidance on deciding between alternative data gathering configurations. Further discussion of the material presented in this paper can be found in O'Sullivan[12].

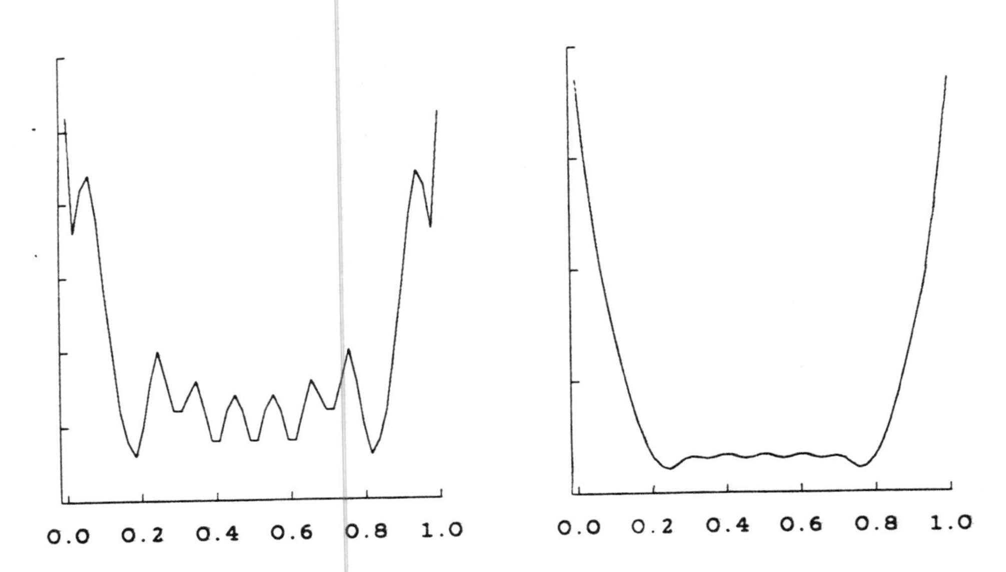

Figure 2.3 : Linearized Maximum Bias (left) and Standard Error (left) for the MOR inversion algorithm applied to the History Matching Problem.

References

1. G. Anger, Inverse and Improperly Posed Problems in Differential Equations: Proceedings of a Conference on Mathematical and Numerical Methods, Akademie-Verlag, Berlin, 1979.

2. G. Backus and F. Gilbert, "Uniqueness in the inversion of inaccurate gross earth data," Philos Trans. Royal Soc. Ser. A. 266 (1970), 123-192.

3. R. D. Cook and J. A. Witmer, "A note on parameter-effects curvature," J. Amer Statist. Assoc. 80 (1985), 872-878.

4. R. L. Cooley, "Incorporation of prior information on parameters into nonlinear regression ground-water water flow models, 1 Theory," Water Resour. Res. 18 (1982), 965-976.

5. D. D. Cox and F. O'Sullivan, "Analysis of penalized likelihood type estimators wih application to generalized smoothing in Sobolev Spaces," Tech. Rep. No. 51, Statistics Dept., University of California, Berkeley, 1985.

6. J. Hadamard, Lectures on Cauchy's Problem, Yale University Press, New Haven, 1923.

7. C. Kravaris and J. H. Seinfeld, "Identification of parameters in distributed parameter systems by regularization," SIAM J. Control and Optimization 23 (1985), 217-241.

8. J. L. Lions, Optimal Control of Systems Governed by Partial Differential Equations, Springer-Verlag, Berlin, 1971.

9. D. W. McLaughlin, Inverse Problems: Proceedings of a Symposium in Applied Mathematics, American Mathematical Society, 1983.

10. S. P. Neuman and S. Yakowitz, "A statistical approach to the inverse problem of aquifer hydrology 1. Theory," Water Resour. Res. 15 (1979), 845-860.

11. D. Nychka, G. Wahba, S. Goldfarb, and T. Pugh, "Cross-validated spline methods for the estimation of three-dimensional tumor size distributions from observations on two dimensional cross sections," J. Amer. Statist. Assoc. 79 (1984), 832-846.

12. F. O'Sullivan, "A practical perspective on ill-posed inverse problems: A review with some new developments," J. Statist. Science (under revision), (1986).

13. L. E. Payne, Improperly Posed Problems in Partial Differential Equations, Regional Conference Series in Applied Mathematics, SIAM, 1975.

14. D. M. Peaceman, Fundamentals of Numerical Reservoir Simulation, Elsevier Scientific Publishing Company, Amsterdam, 1977.

DEPARTMENT OF STATISTICS
UNIVERSITY OF CALIFORNIA, BERKELEY
BERKELEY, CALIFORNIA 94720.

Contemporary Mathematics
Volume **59**, 1986

Harmonic Splines in Geomagnetism

ROBERT L. PARKER

ABSTRACT. The geophysical problem is discussed of predicting the magnetic field of the Earth at the core given measurements made in space by artificial satellites. The problem is not merely one of interpolation of a function in three dimensions since there are severe additional constraints on the class of all possible magnetic fields imposed by the physics. The constraints can be built into a Hilbert space formulation; the modeling problem is ill-posed and the norm of the space provides a natural means for regularization. The smooth basis functions generated by the representers are called harmonic splines. The very large volume of observations makes the direct application of the theory computationally impractical so that an approximation to the harmonic spline solution is usually sought.

1. Introduction. This contribution describes the application of a nonparametric estimation technique to an important problem in Earth sciences. Most of the mathematical ideas will be very familiar to any one working in the field of function estimation, although through ignorance of the literature my colleagues and I reinvented a number of them for this problem. Some new considerations arise from the special needs of the geophysicist who must deal with large volumes of data. In the following, I have gone to some length to provide the scientific context for the benefit of the mathematician who can not be expected to know the background and who may be interested to learn of a new application of nonparametric estimation.

The magnetic field measured at the surface of the Earth consists of contributions from a variety of sources. The most important one,

1980 *Mathematics Subject Classification* (1985 *Revision*). 62G99, 86-08, 86A25.

0271-4132/86 $1.00 + $.25 per page

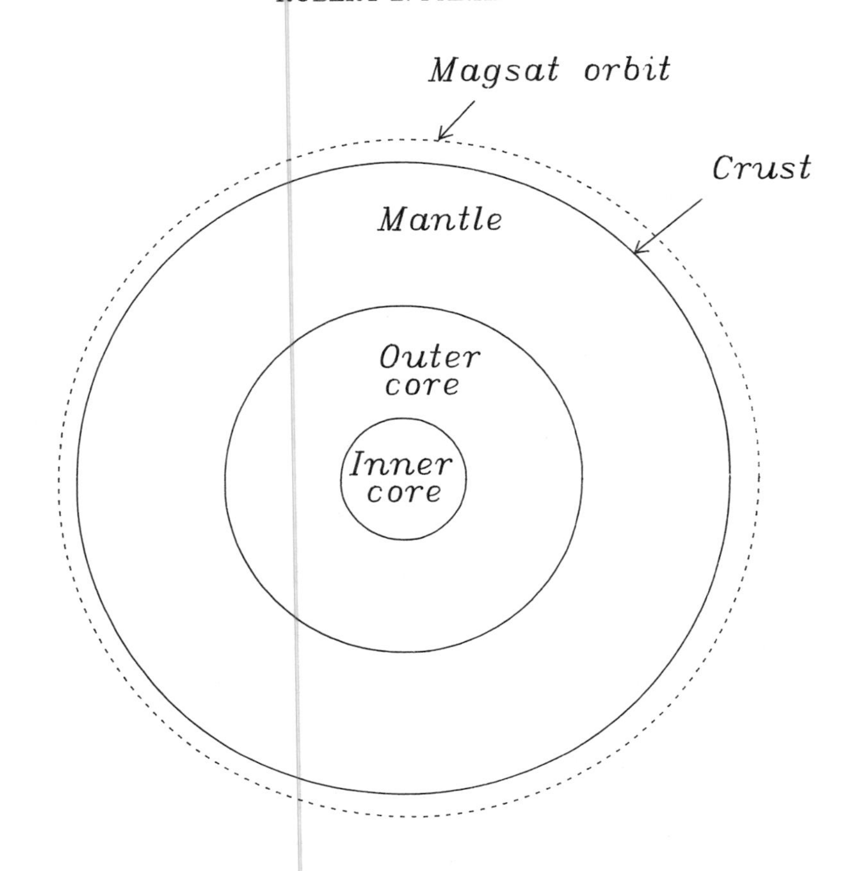

FIGURE 1. A cross section through the Earth showing the principal geophysical divisions. A typical Magsat orbit is also shown. All features are approximately to scale; the Earth's radius is about 6370 km.

comprising over 90 percent of the field in energy, is the geomagnetic dynamo in the Earth's outer core, where motions of a metallic fluid generate electric currents that maintain themselves against resistive loss (see Figure 1 for sketch of a cross section through the Earth showing the major divisions). Geophysicists are not certain what powers the dynamo: there are several candidates, including heat from radioactive decay or stirring by the release of latent heat of solidification. Nor are the details known of how the motions in the core produce the currents, although it is firmly established that such a generation process is possible in principle (Moffatt, 1978). The geomagnetic field is constantly changing on a time scale of five years or more and this is believed to reflect the varying pattern of fluid flow in the core. Under the assumption that the electrical conductivity of the core is large in a precisely defined sense (Roberts and Scott, 1965) the fluid velocity at the surface of the outer core can be derived from the magnetic field and its rate of change with time on the curves where the core field is exactly horizontal (Backus, 1968); with additional assumptions it may be possible to find the velocity on other sets of points from a knowledge of the field at the core (Whaler, 1980; LeMouel and Backus, 1985). Some of the assumptions needed are amenable to test, again by examining the magnetic field at the core (Bloxham and Gubbins, 1986). These brief remarks may indicate why the magnetic field on the surface of the outer core is of considerable interest to one segment of the geophysical community.

Naturally the magnetic field cannot be measured on the the core, which is over 2800 km deep. Measurements at the Earth's surface have been made for nearly 150 years in increasing numbers. There are about 170 geomagnetic observatories recording today but their geographical distribution is very uneven — a large fraction of the stations are in Europe. Poor station distribution is unavoidable in any case because permanent observatories must be on land and there are large expanses of ocean without suitable islands. Since 1965 coverage has been dramatically extended by observations made by artificial satellites; the first measurements were of the field intensity because of the expense of orienting the satellite accurately, but intensity data alone on a spherical surface do not uniquely specify an interior field (Backus, 1970). In 1980 Magsat provided the first virtually global picture of the magnetic vector at an altitude of about 400 km and this data set has been the starting point for several recent investigations of the field at the core (Shure et al., 1985; Bloxham and Gubbins 1986).

The problem of estimating the field at the core is not merely one of interpolation in three dimensions. If we neglect relatively weak sources in the crust, mantle and ionosphere, then Maxwell's equations for the slowly varying electromagnetic field allow us to obtain powerful restrictions that any physical field must satisfy. The core is spherical to a perfectly adequate approximation, so we set the origin of coordinates at its center and take the radius to be c. Then in the region $c \leq |\mathbf{r}| < \infty$ where $\mathbf{r} \in \mathbb{R}^3$ there is a scalar potential Ω such that

$$\mathbf{B} = -\nabla\Omega \tag{1}$$

where $\mathbf{B} \in \mathbb{R}^3$ is the magnetic field. These days magnetic field is measured in units of tesla; at the Earth's surface $|\mathbf{B}|$ varies between 30 to 60 microteslas or μT. Moreover, Ω is harmonic, that is obeys

$$\nabla^2\Omega = 0 \tag{2}$$

and

$$\Omega(\mathbf{r}) = O|\mathbf{r}|^{-2}, \quad |\mathbf{r}| \to \infty \tag{3}$$

These additional properties put strong constraints on **B**. For example, **B** is analytic at every point $|\mathbf{r}| > c$. Also, precise knowledge of the normal component of **B** on any smooth surface enclosing the core is sufficient to determine **B** outside the surface, by solution of (2) with a Neumann boundary condition, and inside it down to the core, by analytic continuation. Our task is to combine measurements made by the satellite with the constraints of physics to construct a plausible model field consistent with both and then to evaluate the field at the surface of the core. With ideal data the core field is uniquely specified, but the problem of construction is ill-posed because of the need for analytic continuation. In the next section we give a representation of the geomagnetic field that has the physical restrictions built into it, together with a natural mechanism for regularizing the instability.

2. A Hilbert Space Formulation. The following is a sketch of the development described by Shure, Parker and Backus (1982). It has been known since Gauss that any magnetic field with the properties (1) (2) and (3) possesses an expansion for its potential in a spherical harmonic series:

$$\Omega(r, \theta, \phi) = c \sum_{l=1}^{\infty} \sum_{m=-l}^{l} b_l^m \left(\frac{c}{r}\right)^{l+1} Y_l^m(\theta, \phi) \tag{4}$$

Here $r = |\mathbf{r}|$ and θ and ϕ are angles in a spherical polar coordinate system; the functions $Y_l^m(\theta, \phi)$ are the spherical harmonics (they are the complex valued, complete orthogonal eigenfunctions of the Beltrami operator, the Laplacian confined to the surface of the unit sphere; see Sansone, 1977); the expansion coefficients in this representation are the complex numbers b_l^m, $l = 1, 2, 3, \cdots$, $|m| \le l$. The absence of the $l = 0$ term is a consequence of the nonexistence or extreme rarity of magnetic monopoles.

We may identify every magnetic field of interest to us with an ordered sequence $\{b_1^{-1}, b_1^0, b_1^1, b_2^{-2}, b_2^{-1}, b_2^0, b_2^1, b_2^2, \cdots\}$ and it is convenient to take each such sequence to be an element b of the Hilbert space complex l_2 by introducing the inner product

$$(a, b) = \sum_{l=1}^{\infty} \sum_{m=-l}^{l} F(l)\, a_l^m (b_l^m)^* \tag{5}$$

and associated norm

$$\|b\| = [\sum_{l=1}^{\infty} \sum_{m=-l}^{l} F(l)\, |b_l^m|^2]^{1/2} \tag{6}$$

where $F(l)$ is a positive polynomial of l that is related to the smoothness of the field on the core and which we will discuss in detail shortly; $F(l)$ is always of degree one or higher in our applications. It follows from (1) and (4) that the x component of $\mathbf{B}$ at a point $\mathbf{r} = (r, \theta, \phi)$ is given by

$$\begin{aligned} \hat{\mathbf{x}}\cdot\mathbf{B}(\mathbf{r}) &= -\frac{\partial \Omega}{\partial x} \\ &= \sum_{l=1}^{\infty} \sum_{m=-l}^{l} -b_l^m c \frac{\partial}{\partial x}\left[\left(\frac{c}{r}\right)^{l+1} Y_l^m(\theta, \phi)\right] \\ &= \sum_{l=1}^{\infty} \sum_{m=-l}^{l} F(l) b_l^m X_l^m(\mathbf{r})^* \end{aligned}$$

where

$$X_l^m(\mathbf{r}) = \frac{-c}{F(l)} \frac{\partial}{\partial x}\left[\left(\frac{c}{r}\right)^{l+1} Y_l^m(\theta, \phi)^*\right]$$

There are of course similar expressions for the other Cartesian components of $\mathbf{B}$. Because $|Y_l^m(\theta, \phi)| = O(1)$ for fixed θ, ϕ, the series

$$\sum_{l=1}^{\infty} \sum_{m=-l}^{l} F(l)\, |X_l^m(\mathbf{r})|^2 = \|X(\mathbf{r})\|^2$$

converges if $r > c$ and therefore evaluation of $\hat{\mathbf{x}}\cdot\mathbf{B}$ is a bounded linear functional of b. The series converges on $r = c$ when the degree of $F(l)$ is three or more. From (5) we may write

$$\hat{\mathbf{x}}\cdot\mathbf{B}(\mathbf{r}) = (b, X(\mathbf{r})) \tag{7}$$

where $X(\mathbf{r}) \in l_2$ is the representer for evaluation of the x component of the field at $\mathbf{r}$.

At this point let us examine the way in which the polynomial $F(l)$ in (6) controls the smoothness of the magnetic fields. The following easily verified identities tell the story:

$$\int_C |\mathbf{B}|^2\, dS = c^2 \sum_{l=1}^{\infty} \sum_{m=-l}^{l} (2l + 1)(l + 1)|b_l^m|^2$$

$$\int_C |\nabla_s \hat{\mathbf{r}}\cdot\mathbf{B}|^2\, dS = \sum_{l=1}^{\infty}\sum_{m=-l}^{l} l(l+1)^2(l+½)|b_l^m|^2$$

$$\int_C |\nabla_s^2 \hat{\mathbf{r}}\cdot\mathbf{B}|^2\, dS = c^{-2}\sum_{l=1}^{\infty}\sum_{m=-l}^{l} l^2(l+1)^4|b_l^m|^2$$

In each of these integrals C is the surface of the core; the operator ∇_s is the surface gradient $\nabla - \hat{\mathbf{r}}\partial/\partial r$ and ∇_s^2 is the Beltrami operator. Thus we see that the norms defined by various $F(l)$ measure different degrees of roughness of the magnetic field at the core: for an increase in the order of differentiation there is an increase by two in the degree of the corresponding polynomial $F(l)$. The components of the vector $\mathbf{B}$ at any point are intimately connected to each other because of (1) and (2), so that roughness in the radial component at the core implies comparable roughness of the others.

Certain other functions F generate norms that possess physical meaning. For example, if $F(l) = (l+1)/2\mu_0$, where μ_0 is the permittivity of free space (a defined constant of the MKS system of units equal to $4\pi\times 10^{-7}$), then $\|b\|^2$ is the total electromagnetic field energy stored outside the core. With $F(l) = (l+1)(2l+1)^2(2l+3)/l$ it can be shown (Gubbins 1975) that $\|b\|^2$ is proportional to a component of the electrical heat generation within the core. Although this function is not a polynomial, it can be approximated by a degree three polynomial so that the maximum possible error in the norm is less than 12 percent (Shure et al., 1982). As we shall explain in a moment, the benefit of restricting F to the polynomials is that great computational savings can be realized in practical calculations.

3. Interpolation of the Field. The generalization of (7) describing the complete measurement series made by the satellite is written

$$d_j = (b, g_j), \quad j=1, 2, 3, \cdots N \tag{8}$$

where g_j is one of the representers for a field component at a point $\mathbf{r}_j$ in the orbit where an observation was made. Suppose for the moment that the observations must be satisfied exactly; then we must choose a field from the infinitely many possible ones. By analogy with spline interpolation, we ask for the smoothest field — that is, the element b with the smallest norm, since the norms measure various kinds of roughness on the core surface. It is well known (Luenberger, 1969) that provided the representer g_j are linearly independent (which they are unless two observation points coincide exactly) there is an element satisfying (8) in the form

$$b_0 = \sum_{j=1}^{N} \alpha_j g_j \tag{9}$$

and b_0 is the unique element in l_2 of smallest norm. To find the coefficients we must solve the linear system

$$\Gamma\alpha = d \tag{10}$$

where α, $d \in \mathbb{R}^N$ are vectors of coefficients and data values and Γ is the Gram matrix of representers

$$\Gamma_{jk} = (g_j, g_k)$$

The solution b_0 is the *harmonic spline* field model. The are a number of ways of describing the solution: the list of coefficients α_j is itself not particularly illuminating. For purposes of visualization we usually contour the radial field component on the surface of the core, using the appropriate representer of evaluation. Just as in ordinary spline interpolation, the spline model is smoother than the general element of the space where the initial problem was posed so that the field components on the core exist and are even analytic at every point. (Recall that the cubic spline interpolator on a line has *continuous* second derivative, while in the setting of the original minimization problem, the second derivative need only be in L_2.) Traditionally, geophysical work involving the main geomagnetic field relies upon its spherical harmonic description. Therefore for quantitative work, the first two hundred or so members of the sequence b_l^m are tabulated; the remainder make a negligible contribution for geophysical purposes. There is no analogy to the B-spline basis of ordinary splines (de Boor, 1978).

The computation of the model and its evaluation via (7) demand the explicit calculation of the Gram matrix Γ. From (5) we need to perform a rather formidable infinite sum which from a computational point of view could be very expensive (there are $N(N+1)/2$ different elements to be evaluated and N is sometimes 300 or more). We have discovered, however, that if the function $F(l)$ is carefully chosen, the inner product can be expressed in terms of elementary functions in closed form. While the expressions are by no means simple ones, they represent an important economy for practical work. For details the reader is referred to Shure et al. (1982 and 1985). For certain norms, for example the one associated with the second integral in our list above, closed form evaluation involves elliptic integrals which are themselves expensive to compute. We chose instead to modify the polynomial slightly so that the norm could no longer be identified with the two-norm of a differential form on the core, but the new norm is topologically equivalent to the old and still describes roughness with the same differential degree.

4. Regularization. We have so far assumed that an exact match to the observations is desirable, but this is not so. While the dynamo-generated portion is by far the largest part of the field measured by the satellite there are other contributions that must be regarded as noise for our purposes. Second in importance are the fields from the magnetic rocks of the crust which give rise to easily measurable signals; the crust is a thin surface shell never more than 40 km thick (see Figure 1). Geological models of the crust (Meyer et al., 1983) give us an estimate of the noise variance and moreover suggest that the noise is reasonably uncorrelated if observation samples are more than 400 km apart (the satellite altitude) as they generally are. Other noise signals are caused by currents flowing in the magnetosphere, a zone about 70,000 km in diameter surrounding the Earth in which energetic protons and electrons are trapped. Magnetospheric noise can be minimized by selecting data for analysis from magnetically quiet periods. The ionosphere is an electric-current-carrying region at about 100 km altitude; the major currents flow on the sun-lit hemisphere. The orbit of Magsat was designed to minimize ionospheric effects by its placement in the dusk-dawn plane.

Even tiny errors are magnified by the so called 'downward continuation' of the magnetic field from its measurement region to the core's surface. This is illustrated with a very simple example: suppose the radial component of $\mathbf{B}$ is measured on a spherical surface radius C and that the measurements are contaminated with a white noise n with covariance function $\sigma^2\delta(\hat{\mathbf{r}})$. From (1) and (4) the radial field is

$$B_r(C,\,\theta,\,\phi) = -\frac{\partial\Omega}{\partial r} + n$$

$$= \sum_{l=1}^{\infty}\sum_{m=-l}^{l} b_l^m\, l\left(\frac{c}{C}\right)^{l+2} Y_l^m(\theta,\,\phi) + n$$

It follows from the orthogonality of the spherical harmonics under the inner product over the unit sphere S^2

$$\langle f,\, g\rangle = \int_{S^2} f(\theta,\,\phi)g(\theta,\,\phi)^*\,\sin\theta\; d\theta d\phi$$

that estimates of the coefficients b_l^m are furnished by

$$\hat{b}_l^m = \langle B_r,\, y_l^m\rangle$$

where $y_l^m(\theta,\phi) = (C/c)^{l+2} Y_l^m(\theta, \phi)/l$ and the variance of $\hat{b}_l^m$ $(C/c)^{2l+4}\sigma^2$. When the measurement surface is at or above the Earth's surface $C/c \geq 1.78$ and so the variance grows exponentially with increasing spherical harmonic degree l. From considerations of energy (see the discussion of various norms above) every physically realizable field must be associated with a sequence of coefficients that decreases with l and thus the estimates become overwhelmed by noise at the shorter wavelengths — recall that $2\pi/l$ corresponds roughly to the wavelength of oscillation of the associated eigenfunction $Y_l^m(\theta, \phi)$ on the unit sphere. Even though the idealized problem differs from the practical one in several ways (e.g. measurements are not made on a sphere of constant radius and they are only finitely many data samples) it demonstrates the need for a regularization process to suppress the random short wavelength components or, equivalently, to smooth the estimated core fields and coefficients.

There are two slightly different ways of deriving the solutions. The standard mathematical approach to regularization asks for the element of l_2 that minimizes the functional

$$\lambda \|b\|^2 + \sum_{j=1}^{N} [d_j - (b, g_j)]^2 \tag{11}$$

where λ is a smoothing parameter. We have already seen how $\|b\|$ is a roughness measure of the core field so that the functional penalizes some combination of model misfit and model roughness. Here the noise is treated as a set of independent, identically distributed random variables. In the practical problem it is necessary to allow the variances to vary with position so that the random variables must be standardized and the representers scaled appropriately; we shall skip this trivial complication in our discussion. As with interpolation, the minimizing element has the form of (9), a linear combination of representers, but in place of (10) the expansion coefficients obey

$$(\lambda I + \Gamma)\alpha = d \tag{12}$$

Mathematical interest centers on the proper choice of λ as N grows so that consistent estimates are obtained and the optimal convergence rate is secured.

In the geophysical derivation we arrive at (11) by another route. With actual data the number of observation N is fixed though large. We ask for the smoothest model in the sense of $\|b\|$ such that the observations are satisfied according to a standard misfit criterion. Then (11) and (12) are obtained and λ is the Lagrange multiplier for the optimization problem. We determine λ by finding which member of the one-dimensional manifold of elements making (11) stationary obeys

$$\sum_{j=1}^{N} [d_j - (b, g_j)]^2 = \sigma^2 N \tag{13}$$

where is σ^2 is the standardized variance of each measurement. A major difficulty arises from the very large number of observations available. The original set of magnetic readings constitutes many hundreds of thousands of numbers. These are usually edited to a list of several thousand, as evenly distributed as possible in space and selected to avoid periods of magnetic disturbance in the magnetosphere and ionosphere. Typically $N = 4000$ or so and then the linear system (12) is unmanagably large.

The traditional solution to the core-field modeling problem is the one invented by Gauss: equation (4) is replaced by one in which the highest l (called the degree of the spherical harmonic) is L and an ordinary least squares fit is performed between the predictions of the truncated spherical harmonic series and the large data set. The computational advantage is that the number of computer operations needed to obtain a model is about $NL^4/8$ and $L^2 << N$ (typically N is less than 15) instead of growing like $N^3/6$ as with (12). Of course, the disadvantage of traditional approach is that smoothing is accomplished by an arbitrary truncation of a series and, at the same level of misfit, this results in an appreciably rougher core field (Shure et al., 1985); some of the noise in the data has been unnecessarily ascribed to the main field. If the harmonic spline approach is to displace the ordinary least squares recipe, something must be done to improve its computational efficiency; to meet this need the idea of a 'depleted basis' for the harmonic splines was invented (Parker and Shure, 1982).

The maximal smoothness of the core field seems to be at odds with very large number of parameters seemingly needed for its description. In (11) the set of elements over which the minimization is performed is all of l_2 and, because of the Projection Theorem for Hilbert spaces, it turns out that the minimizing element lies in the subspace dimension N spanned by the representers g_j. Suppose the minimization is taken over a smaller subspace still, one constructed from a basis of $M < N$ representers. The representers g_j are each associated with a position in ordinary space, for example $\mathbf{r}$ in the case of $X(\mathbf{r})$; if the M representers in the depleted basis are chosen to have a reasonably uniform geographic distribution with respect to these locations, it seems plausible that a good approximation to the true roughness-minimizing solution could be obtained in the smaller subspace. The size of the associated linear system is naturally greatly reduced by this device. It is obvious that there other are possible choices for the subspace over which (11) may be minimized; Gubbins and Bloxham (1985) employed a subspace with a finite number of spherical harmonics as basis elements.

If we view the problem as one of approximating the optimally smooth $b \in l_2$ by an element of a small subspace, it is easy to write down a theory to give a bound on the error in such an approximation, but the numerical calculations to take advantage of this theory are even more burdensome for large N than the one the depleted basis method basis was designed to avoid. In small test problems (Parker and Shure, 1982) where the optimally smooth b can be found exactly, we observe very good approximation when M is a small fraction of N. We calculate a sequence of models with M increasing, tabulating $\|b\|$. It is found, after a certain point, the roughness ceases to decrease significantly and then in the test models we find the model of smallest norm is well approximated. This heuristic strategy was adopted for the large data set (Shure et al., 1985); here $N = 4180$ and in the final harmonic spline model, $M = 180$.

In Figure 2 we compare this harmonic spline core field derived from Magsat data with another obtained by least squares fitting, GSFC 9/80 (Langel et al., 1982). The degree of truncation L for the upper model is 13. The agreement with the observations is almost exactly the same in both cases. Contours of the radial field are plotted at intervals of 400 μT show how much smoother the harmonic spline model is. As mentioned in the introduction, the null flux curves where $B_r = 0$ are of special significance to theories of the fluid motion; this contour is shown heavy and $B_r < 0$ regions have dashed contours. There are two more closed null flux curves in the model obtained by least squares than in the harmonic spline model.

5. Mathematical Issues. In the geophysical application of the harmonic spline method a number of important mathematical issues have been ignored. For example, under what conditions and how rapidly does the procedure converge to the true field in the limit of large N and what is the proper choice of smoothing parameter λ? It has become clear (see, for example the contributions of John Rice and David Donoho in this volume) that the geophysicists' choice for λ is not a good one in the limit. Considerable progress has been made recently on closely related idealized systems by Hobbs (1985). He shows that if sufficient smoothness of the true core field is assumed, convergence can be guaranteed, but at a very slow rate even when the best choice of λ is made.

It would be very useful for practical solutions involving the depleted

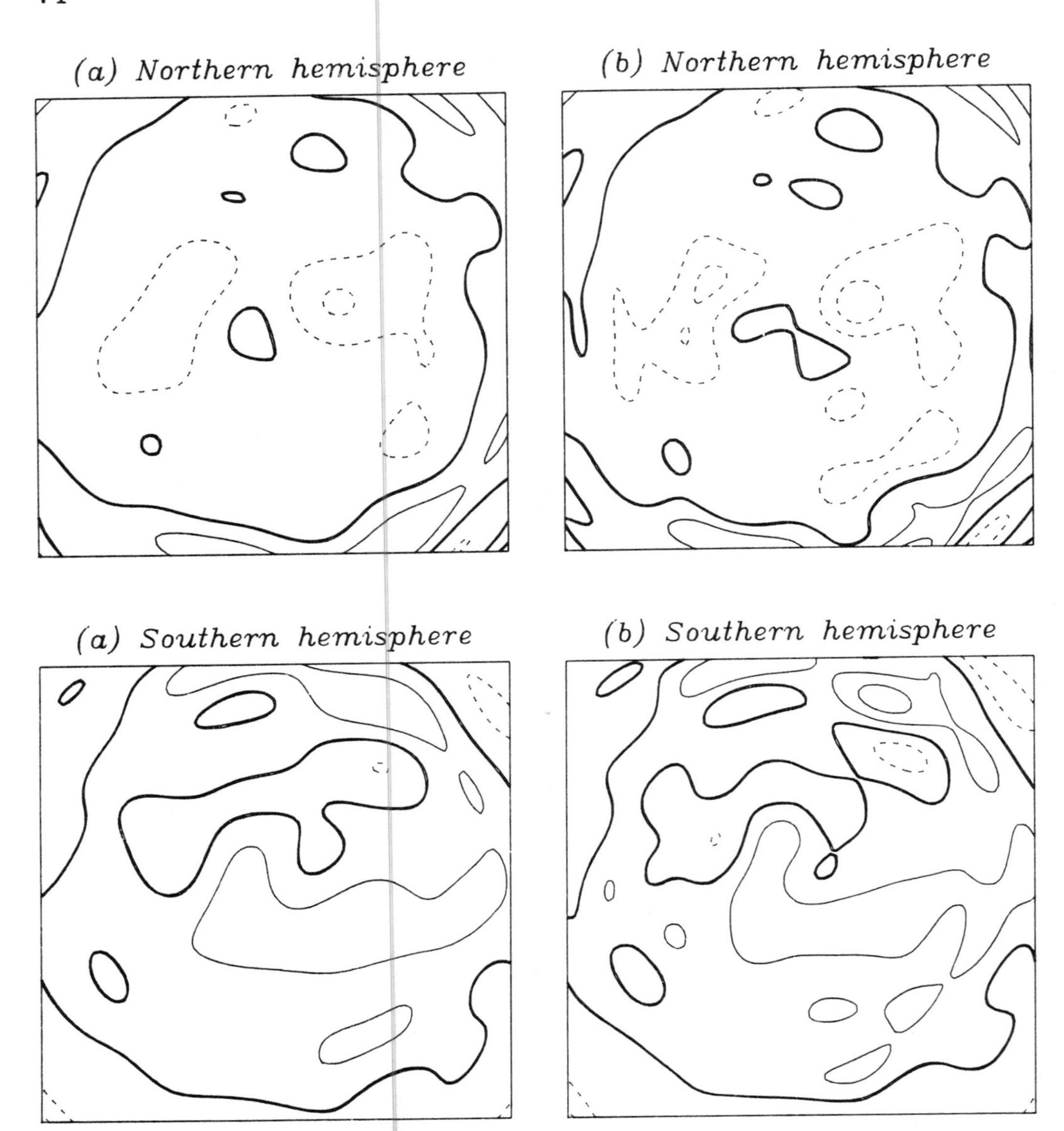

FIGURE 2. North and South polar views of the radial component of the magnetic field at the core seen in a Lambert equal-area projection according to two models based on Magsat data contoured at 400 μT intervals: (*a*) GSFC 9/80, a truncated spherical harmonic series; (*b*) a harmonic spline solution, HSP 80. The Greenwhich meridian bisects each frame vertically and the Western hemisphere is on the left; the equator is the inscribed circle of each square.

basis to have a proper theory for the error of the approximation to the optimal smooth model. An approach to this question is suggested by the asymptotic theory of ordinary smoothing splines given by Silverman (1984). When smoothing is strong and the number of data is large he shows how to derive an excellent approximation for the linear kernel that maps the measured values into the smooth curve and that approximation can be obtained without any heavy computations — all that is needed is the parameter λ. If the elements in the depleted basis are capable of giving a good approximation to the asymptotic kernel, they must be able to approximate the smooth model because the kernel is just the smoothest approximation to an impulse function in the data, which is the worst case. These ideas have been used for a depleted basis theory of ordinary cubic splines (Constable and Parker, 1986) and they should generalize without much difficulty to harmonic splines.

BIBLIOGRAPHY

Backus, G. E., 1968, *Kinematics of geomagnetic secular variation in a perfectly conducting core,* Philos. Trans. R. Soc. London, Ser A. **263**, 239-266.

Backus, G. E., 1970, *Nonuniqueness of the external geomagnetic field determined by intensity measurements,* J. Geophys. Res., **75**, 6337-6341.

Backus, G. and J. L. LeMouel, 1985, *The region on the core-mantle boundary where a geostrophic velocity field can be determined from frozen-in magnetic data,* Trans. Amer. Geophys. Union, **66**, 878.

Bloxham, J. and D. Gubbins, 1986, *Geomagnetic field analysis — IV. Testing the frozen-flux hypothesis,* Geophys. J. R. astr. Soc., **84**, 139-152.

Constable, C. G. and R. L. Parker, 1986, *Smoothing, splines and smoothing splines,* Submitted to SIAM J. Sci. Stat. Comp.

de Boor, C., 1978, *A Practical Guide to Splines,* Springer-Verlag, New York.

Gubbins, D., 1975, *Can the Earth's magnetic field be sustained by core oscillations?,* Geophys. Res. Lett., **2**, 409-412.

Gubbins, D. and J. Bloxham, 1985, *Geomagnetic field analysis — III. Magnetic fields on the core-mantle boundary,* Geophys. J. R. astr. Soc., **80**, 695-713.

Hobbs, S. L., 1985, *Statistical Properties of a Nonparametric Regression Function on S^2,* PhD Dissertation, University of California, San Diego.

Langel, R. A., R. H. Estes and G. D. Mead., 1982, *Some new methods in geomagnetic field modeling applied to the 1960-1980 epoch,* J. Geomag. Geoelec., **34**, 327-349.

Luenberger, D. G., 1969, *Optimization by Vector Space Methods,* John Wiley and Sons, New York.

Meyer, J., J.-H. Hufen, M. Siebert and A. Hahn, 1983, *Investigations of the internal geomagnetic field by means of a global model of the earth's crust,* J. Geophys., **52**, 71-84.

Moffatt, H. K., 1978, *Magnetic Field Generation in Electrically Conducting Fluids,* Cambridge University Press, Cambridge.

Parker, R. L. and L. Shure, 1982, *Efficient modeling of the earth's magnetic field with harmonic splines,* Geophys. Res. Lett., **9**, 812-815.

Roberts, P. H. and S. Scott, 1965, *On the analysis of the secular variation I. A hydromagnetic constraint,* J. Geomag. Geoelec., **17**, 137-151.

Sansone, G., 1977, *Orthogonal Functions,* Krieger Pub. Co., New York.

Shure, L., R. L. Parker and G. E. Backus, 1982, *Harmonic splines for geomagnetic modelling,* Phys. Earth Planet. Int., **28**, 215-229.

Shure, L., R. L. Parker and R. A. Langel, 1985, *A preliminary harmonic spline model from Magsat data,* J. Geophys. Res., **90**, 11505-11512.

Silverman, B. W., 1984, *Spline smoothing: the equivalent kernel method,* Annals. Stats., **12(3)**, 898-916.

Whaler, K. A., 1980, *Does the whole of the Earth's core convect?,* Nature, **287**, 528-530.

INSTITUTE OF GEOPHYSICS AND PLANETARY PHYSICS
A-025 UNIVERSITY OF CALIFORNIA, SAN DIEGO
LA JOLLA, CALIFORNIA 92093.

Contemporary Mathematics
Volume 59, 1986

PROBLEMS IN ESTIMATING THE ANOMALOUS GRAVITY POTENTIAL OF THE EARTH FROM DISCRETE SPATIAL DATA

K.P. Schwarz

ABSTRACT. A major task of geodesy is the determination of a unique and coherent representation of the gravity field at the Earth's surface and in space. This representation has to be estimated from discrete and noisy data which are heterogeneous in nature and do not cover the surface of the Earth in a regular fashion. The paper discusses current methods to determine the gravity potential and its linear functionals and emphasizes some of the unsolved problems.

After a brief review of gravity field fundamentals, the basic solution approaches are introduced and some of the data related problems are discussed. Recent developments in the solution of geodetic boundary value problems are presented and the alternative approach, approximation in a reproducing kernel Hilbert space, is outlined. In both areas, significant theoretical problems remain to be solved. A brief discussion of the instabilities encountered when using airborne and satellite gravity data, and of the difficulties with the use of regularization techniques in this case, concludes the paper.

1. INTRODUCTION

One of the major tasks of geodesy is the determination of a unique and coherent representation of the gravity field from discrete, noisy data taken at the Earth's surface or in space. This representation is needed to model geodetic measurements, to predict perturbations of satellite orbits, to determine global ocean circulation patterns, to assist global geophysics, and to support oil and mineral exploration.

The representation is usually given in terms of the gravity potential $W(x,y,z)$ which is split into a normal part $U(x,y,z)$ and an anomalous part $T(x,y,z)$. The normal part U is given by the potential of an equipotential ellipsoid with the same mass and angular velocity as the Earth and is used as a convenient gravity and geometric reference. The problem thus consists in estimating the anomalous potential T as the spatial function that describes deviations from this reference.

The solution of this problem has to consider a number of constraints which are either due to the physics of the problem or to the way in which the data are taken. Constraints of the first type are the harmonicity of T outside the Earth's surface and its regularity at infinity; the need for combining heterogeneous functionals of T in the solution; the ill-posedness of the problem when combining data taken at satellite and aircraft altitudes with data taken at the surface of the earth. Constraints of the second type are the absence of a uniform global data set of sufficient density, large variations in the measurement noise, the enormous amount of data to be handled, and the fact that a regular data distribution is the exception rather than the rule. In addition, the accuracy of the results as a function of data accuracy and distribution is usually required. Thus, data related problems are very real in geodesy and determine to a large extent the implementation of specific solution approaches.

Felix Klein (1928), in his classical lectures on 'Elementary Mathematics from an Advanced Standpoint', characterized *'geodesy as that part of geometry where the idea of mathematical approximation has found its clearest and most consequent expression. One studies consistently on the one hand the accuracy of the observation and on the other the accuracy of the ensuing results'*. This data dependence is still very much a characteristic of the geodetic approach and the resulting mix of approximation theory, statistical estimation, and data handling problems, although a horror to purists, is certainly a challenge to those working in the field.

2. THE EARTH'S GRAVITY FIELD

This section summarizes some basic properties of the earth's gravity field and introduces standard notation used in geodesy. It closely follows Moritz (1980) which should be consulted for all details.

The gravitational potential of the earth is expressed by

$$(1) \qquad V(P) = V(x,y,z) = G \iiint_E \frac{\rho(Q)}{\ell} \, dv_Q$$

where P is a point outside the Earth's surface, Q is the centre of the volume element dv_Q and is variable within the Earth's body E, ℓ is the distance between P and Q, $\rho(Q)$ is the mass density at Q considered constant for dv_Q, and G is the Newtonian gravitational constant. The usual assumptions that the earth is a rigid body without atmosphere and without temporal variations will be made. They introduce errors of the relative order of 10^{-6} and 10^{-7}, respectively, which are negligible for the problem considered here.

In general, an earth-fixed rectangular coordinate system will be used which is defined in the following way : the origin is at the Earth's centre of mass; the z-axis coincides with the mean axis of rotation and is thus normal to the mean equatorial plane which is the xy-plane; within this plane the mean meridian of Greenwich determines the direction of the x-axis and the definition of a right-handed system the y-axis. The mean spin axis and the mean Greenwich meridian plane are used to get a time-independent definition.

The gravity potential of the Earth is expressed by

$$W(P) = V(P) + V_c(P) \tag{2a}$$

where

$$V_c(P) = \frac{1}{2}\omega^2 (x^2 + y^2) \tag{2b}$$

is the potential of the centrifugal force and ω the angular velocity of the Earth's rotation. Again, ω can be considered as constant if an error of the relative order of 10^{-7} is admissable. An equipotential surface is defined by

$$W(P) = \text{const.} \tag{3}$$

The equipotential surface most commonly used as a reference is the geoid which, somewhat loosely, can be described as the idealized surface of the oceans.

The gravity vector $\vec{g}$ is the gradient of W, i.e.

$$\vec{g} = \text{grad } W(P) = \begin{pmatrix} W_x \\ W_y \\ W_z \end{pmatrix} \tag{4a}$$

where the components W_x, W_y, W_z are the partial derivatives with respect to x,y and z. Of particular interest is the magnitude of this vector because it is a measurable quantity. It is usually denoted by

$$g = |\vec{g}| \ . \tag{4b}$$

The direction of $\vec{g}$ is defined by the unit vector $\vec{n}$

$$\vec{n} = \begin{pmatrix} \cos\Phi \cos\Lambda \\ \cos\Phi \sin\Lambda \\ \sin\Phi \end{pmatrix} \tag{4c}$$

where the astronomical latitude Φ and the astronomical longitude Λ are again measurable quantities.

The second-order derivatives of W or the gravity gradients are given by the tensor

$$(5a)\qquad W_{ij} = \begin{pmatrix} W_{xx} & W_{xy} & W_{xz} \\ W_{yz} & W_{yy} & W_{yz} \\ W_{zx} & W_{zy} & W_{zz} \end{pmatrix}$$

whose diagonal components satisfy

$$(5b)\qquad \begin{aligned} \Delta W &= 2\omega^2 \quad \text{outside the Earth's surface, and} \\ \Delta W &= -4\pi G\rho + 2\omega^2 \quad \text{inside the Earth's surface,} \end{aligned}$$

where

$$\Delta W = W_{xx} + W_{yy} + W_{zz} \quad .$$

The corresponding components of the gravitational gradient tensor V_{ij} satisfy Laplace's equation $\Delta V = 0$ in the first case, and Poisson's equation $\Delta V = -4\pi G\rho$ in the second. Because of (5b) and the symmetry of the tensor, W_{ij} has five independent components which are again measurable.

The normal gravity field is defined by

$$(6)\qquad U(P) = \bar{V}(P) + V_c$$

where $\bar{V}(P)$ is the potential of the homogeneous ellipsoid which is the best approximation to the actual geoid. U(P) can be determined by fixing four parameters and by postulating that the resulting ellipsoid is an equipotential surface for U, i.e.

$$(7)\qquad U(x,y,z) = U_o = \text{const.}$$

A typical set of parameters which can be derived from measurements is

a,b ... semiaxes of the ellipsoid

ω ... angular velocity of Earth rotation

M ... mass of the Earth.

The deviation between the actual gravity potential W and the normal potential U is small, as is the deviation between the geoid and a best-fitting ellipsoid, which nowhere exceeds 100 m. Since U(Q) can be expressed by closed formulas, it is often used as a convenient first approximation of W(P). Similarly, the normal gravity vector

(8) $\vec{\gamma}(Q) = \text{grad } U(Q)$

is used as a first approximation for $\vec{g}(P)$. Its magnitude can again be expressed by a closed formula while its direction is normal to the ellipsoid.

Thus, the anomalous gravity potential T is defined as the difference

(9) $T(P) = W(P) - U(Q)$,

having the anomalous gravity vector $\Delta\vec{g}$

(10a) $\Delta\vec{g} = \vec{g} - \vec{\gamma}$

with components

$$(10b) \qquad \Delta\vec{g} = \begin{pmatrix} \xi \\ \eta \\ \Delta g \end{pmatrix} = \begin{pmatrix} \Phi - \phi \\ (\Lambda - \lambda)\cos\phi \\ |\vec{g} - \vec{\gamma}| \end{pmatrix}$$

where ϕ and λ are the ellipsoidal latitude and longitude, respectively.

The components ξ and η are called deflections of the vertical and do usually not exceed 10 - 20 arcseconds; Δg is called gravity anomaly. T and its derivatives will in the following be used for the problem definition. In linear approximation, the components of $\Delta\vec{g}$ depend on T in the following way

$$\xi = -\frac{1}{\gamma}\frac{\partial T}{\partial u}$$

$$\eta = -\frac{1}{\gamma}\frac{\partial T}{\partial v}$$

$$\Delta g = -\frac{\partial T}{\partial r} - \frac{2}{r}T$$

where u, v are local Cartesian coordinates in the tangent plane to the geoid at P, and r is the distance between P and the Earth's centre of mass. Similar simple relations can be established for the anomalous gravity gradients T_{ij}.

3. SOLUTION APPROACHES AND DATA PROBLEMS

Stated in a somewhat loose fashion, gravity field approximation is the estimation of the anomalous potential T in R^3 outside the Earth's surface S from discrete and noisy data given on S and in R^3. Two major approaches to the solution of this problem have been proposed. One considers only data on S, the other data in R^3 and on S. Both will be briefly reviewed in this chapter with specific emphasis on their data dependence. A discussion of the mathematical problems inherent in each approach will be given in the next chapters.

In the first approach, a boundary value problem for the gravity anomalies is formulated. Its solution is approximated either by a series expansion into solid spherical harmonics or by integrals. In spherical approximation, i.e. accepting errors of the relative order of the Earth's flattening in T, the series expansion of the linearized problem results in

$$T(P) = \sum_{n=2}^{\infty} \left(\frac{a}{r}\right)^n \sum_{m=0}^{n} P_{nm}(\cos\theta)\{J_{nm}\cos m\lambda + K_{nm}\sin m\lambda\} \tag{11}$$

where P_{nm} are the associated Legendre functions, J_{nm} and K_{nm} are spherical harmonic coefficients, a is the semimajor axis of the reference ellipsoid, (r,θ,λ) are spherical coordinates. Summation for n starts at 2 in order to constrain the origin of the coordinate system to the centre of mass and the coordinate axes to the three principal axes of inertia of the Earth. These physical constraints are used to obtain a unique solution; formally T must be invariant to translations of the coordinate system. Uniform convergence of the series to the surface of the Earth has not been shown; for a discussion see Moritz (1980). The downward continuation problem is apparent in the factor $(a/r)^n$. For satellite altitudes the quotient (a/r) will always be smaller than 1. High frequency information at satellite altitudes is therefore strongly damped. Deriving a solution at the Earth's surface from these data by analytical continuation is an unstable process.

The solution of the boundary value problem by integral equations leads, for a known spherical boundary, to Stokes' integral and, for an unknown boundary surface, to Molodensky's problem and the resulting integral series. In its simplest form, the solution is

$$T(P) = \frac{M}{4\pi} \iint_{\sigma} \Delta g_Q \, S(\psi_{PQ}) \, d\sigma_Q + T_o + T_1(\theta,\lambda) \tag{12}$$

where M is the mean radius of the Earth, Δg is the gravity anomaly function on the sphere σ with radius M, $S(\psi)$ is Stokes function – an analytical summation of an infinite series –, ψ is the spherical distance between P and Q, and T_o and T_1 are again the zero and first-order terms which are often fixed in the same manner as above. It should be noted that T is determined from one data type only, the gravity anomalies Δg. The relation between T at the boundary surface S and T_e in the space outside S can be obtained by solving Dirichlet's problem for the exterior of the sphere. This leads to Poisson's integral

$$T_e(P) = \frac{M(r_P^2 - M^2)}{4\pi} \iint_{\sigma} \frac{T}{\ell^3} \, d\sigma \tag{13}$$

where ℓ is the distance between the surface point and the exterior point P and r_P is the distance between the centre of mass of the Earth and P. The downward continuation problem, i.e. finding T from discrete measurements T_e, is again an unstable problem. It plays a role in airborne gravimetry and gradiometry and is discussed as a regularization problem in Schwarz (1979).

Boundary value problems are important in geodesy because they define the minimum information needed to construct an approximation of T outside S. However, in terms of data availability, these formulations are not realistic. A more general approach has been proposed in Krarup (1969) who sought the solution in a reproducing kernel Hilbert space. The collocation solution advocated by him is of the general form

$$(14) \qquad T_i = \{L_i K(P,Q)\}^T \{L_j \, L_i \, K(P,Q)\}^{-1} \, \delta b_j$$

where L are bounded evaluation functionals, δb_j are measurements, and K(P,Q) is the reproducing kernel

$$(15) \qquad K(P,Q) = \sum_{n=0}^{\infty} \sigma_n \left(\frac{M^2}{r_P r_Q} \right)^{n+1} P_n(\cos\psi) \; .$$

The terms σ_n are dependent on the choice of the base functions. They are called degree variances for spherical harmonics and describe a kind of decay rate for increasing degree n, for details see Tscherning and Rapp (1974). The advantage of the solution (15) is that it allows to treat all functionals of T as observables and thus opens the way to handle the actual data situation. It also is not restricted to data on the boundary surface but allows the consistent treatment of data on S and in R^3 outside S. The downward continuation is obviously not eliminated - it shows clearly in the the term $(M^2/r_P r_Q)^{n+1}$ in the kernel - but it becomes somewhat more tractable from a numerical point of view. The choice of the norm introduces a certain arbitrariness into the solution, see e.g. Tscherning (1977) for a discussion, and convergence of the solution has not been shown. Recently, Svensson (1983) has raised objections to the use of collocation in mixed data situations because the problem may not be well posed in terms of the mean square norm. Thus, further analysis of this solution approach is needed.

The representations (11), (12), and (14) will now be examined from the data point of view, discussing specifically the observability of T and its functionals and the availability of different data types. The anomalous potential T is not directly observable. It can be derived from orbit perturbations of satellites, from satellite altimeter data, and, in terms of potential differences, from a combination of levelling and gravity measurements. The three first derivatives of T, i.e. the components of gravity anomaly vector (10b) are all observable. The deflections of the vertical are obtained from a combination of astronomical and classical or satellite positioning techniques, the gravity anomaly from direct gravity measurements. Similarly, the five independent components of the gravity gradient tensor will be available from satellite and airborne gradiometer systems.

In terms of availability, the most serious drawback is that none of the above data types is available with a regular global distribution. In addition, the orbital perturbation data, which come closest to a global set, are low pass filtered, reflecting the attenuation of the gravity field at satellite altitudes, and can therefore only be used to extract low frequency information. The satellite altimeter data, because of the measurement technique used, are only available in ocean areas. Similarly, the bulk of gravity measurements is on land. While altimeter data show a regular distribution, the coverage with land gravity data has large gaps, largely coinciding with national boundaries and inaccessible areas. Each of these sets contains between half a million to several million data. Deflection measurements, because of their high costs, are relatively scarce and poorly distributed from a global point of view. Gravity gradiometer data are not yet available but will be extremely important for future work. If the planned satellite gradiometer mission or the alternative satellite to satellite tracking mission gets under way, it will deliver the first truly global data set for the determination of T. Airborne gravity gradiometry, which is expected to be operational in 1987, will contribute greatly to the short wavelength resolution, i.e. to the local representation of the gravity field. For a more detailed discussion of data aspects, see Schwarz (1984).

In order to resolve the coefficients in the series (11) a regular global data set is required. The closest one could get to such a set today is by combining orbital perturbation data, satellite altimetry, and gravity anomaly data. All three data types can be expressed by infinite series of the same spherical harmonic coefficients $\{J_{nm}, K_{nm}\}$ so that, in principle, the solution of a large system of linear equations would provide an approximation to the problem. The difficulty lies in the current numerical capabilities. The number of observations is several millions. The number of coefficients to be determined depends on the degree of resolution wanted. For many local applications wavelengths of about 10 km should be resolved. This corresponds to an expansion of degree and order 4000, i.e. a series with about 16 million coefficients. Such a solution is clearly not feasible. Even a modest resolution obtained from an expansion of degree and order 180 would require the inversion of a matrix of dimension 32 500. This would not only severely tax the capabilities of present supercomputers but also leave considerable doubt about the usefulness of the results.

To overcome these numerical difficulties, the orthogonality of the spherical harmonics is used to determine the coefficients J_{nm}, K_{mn} by integration using the well-known relations

$$(16) \qquad \left\{ \begin{matrix} J_{nm} \\ K_{nm} \end{matrix} \right\} = \frac{1}{4\pi} \iint_{\sigma} f(\theta,\lambda)\, P_{nm}(\cos\theta) \left\{ \begin{matrix} \cos m\lambda \\ \sin m\lambda \end{matrix} \right\} .$$

To get a global $f(\theta,\lambda)$ for the integration, the gravity anomaly values are used to estimate $(1° \times 1°)$ mean gravity anomalies on land and the satellite altimeter data are used to get a similar set in the oceans. The latter requires the solution of an equation of type (12) for Δg, i.e. it leads to an improperly posed problem. A discussion of this problem and some numerical approaches are given in Rummel (1977). A low degree harmonic expansion obtained from satellite perturbations is then combined with the $(1° \times 1°)$ mean anomaly set to obtain the final expansion. For details on the data combination and weighting procedures, see Rapp (1981). At present, expansions of degree and order 180 are available and fast programs have been developed to make efficient use of these coefficient sets, see e.g. Tscherning et al (1983). Alternative approaches using splines have been proposed, see Meissl (1981) and Freeden (1982), but have so far not been applied.

Integrals of type (12) have the obvious advantage that the global data density has not to be uniform. Thus, dense local data can be used close to the integration point while mean values or low-degree series expansions can be used in the more distant zones. Since the kernel function $S(\psi)$ attenuates with increasing ψ, such solutions often allow a better use of the local data than series expansions. Their drawback is the restriction to one data type, gravity anomalies Δg. As has been mentioned before, this data is sufficiently dense only in land areas and the solution of the first-order integral equation (12) is again required to get Δg in the oceans. In the past few years, the solution of the mixed boundary value problem, using Δg on land and T in the oceans, has therefore received considerable attention. It will be briefly discussed in the next chapter.

The collocation solution is, without doubt, the most consistent approach presently available to handle the problem of heterogeneous data. It also is the only method which allows to treat directly discrete noisy measurements with an irregular spatial distribution. Its major drawback from a numerical point of view is the necessity to invert very large matrices. The dimension of these matrices is equal to the number of measurements used which immediately excludes solutions with a global data set. This approach is therefore restricted to local data sets where the long wavelength information is supplied in some other way. Stepwise procedures have been used to somewhat alleviate the problem, see e.g. Moritz (1980), and spectral methods have recently proved very efficient for a regular data distribution. The method has gained wide acceptance in practice although some numerical limitations remain.

4. GEODETIC BOUNDARY VALUE PROBLEMS

The following overview will be rather brief and will be restricted to outlining the major results achieved in recent years. For details, Moritz (1980) or the review paper by Sanso (1981) are recommended.

The simplest geodetic boundary value is that of Stokes which in its current version is usually formulated: Given the gravity anomaly Δg on the surface S_0 of a geocentric ellipsoid, determine T on and outside S_0 such that

$$\Delta T = 0 \quad \text{outside } S_0$$

$$\frac{\partial T}{\partial r} + \frac{2}{M} T + \Delta g = 0 \quad \text{on a sphere with radius } M \tag{17}$$

$$T(r) = 0 \quad \text{as } r \to \infty .$$

The solution of this problem can be written in the form (12). The major assumption in the problem formulation is that Δg is given on the ellipsoid. Since Δg is measured at the surface of the Earth, assumptions about the density of the masses between surface and ellipsoid are necessary. Further assumptions are made in the linearization of the problem and in projecting Δg from the ellipsoid onto the sphere. This projection, usually called spherical approximation, simplifies the problem to a normal derivative problem for a fixed boundary. Errors in T introduced by this assumption are of the order of the ellipsoidal flattening (0.003). Approaches to extend Stokes' problem to the ellipsoid are summarized in Moritz (1980).

To overcome the major objection to this problem formulation, Molodensky in 1945 approached the problem for gravity given at the surface of the Earth. His work stimulated new interest in the field and has, especially in recent years, led to a number of new results. The problem can be stated as follows (Moritz, 1980): Given at all points of the surface S of the Earth the gravity potential W and the gravity vector g, determine the surface S. The problem is thus a free boundary value problem and is of the oblique derivative variety because grad W is not orthogonal to the Earth's surface. Molodensky linearized the problem by introducing a reference potential U and a reference surface S_o close to the actual potential W and the actual surface S. By means of these surfaces the nonlinear free boundary value problem is reduced to a fixed boundary value problem. In contrast to this approach, a fixed boundary value approach for the known surface of the Earth has been advocated by a number of authors, see e.g. Koch (1971). The surface is in this case supposed to be known from satellite positioning, and many of the theoretical difficulties disappear, for a discussion and comparison, see Grafarend et al (1985). The reference surface S_o, called the telluroid, can be defined in a number of ways. Using the notations of Figure 1, two frequently used definitions are

$$\Delta W = W_P - U_Q = 0$$

or

$$\Delta g = g_P - \gamma_P = 0 ,$$

leading to the so-called Marussi telluroid and gravimetric telluroid, respectively.

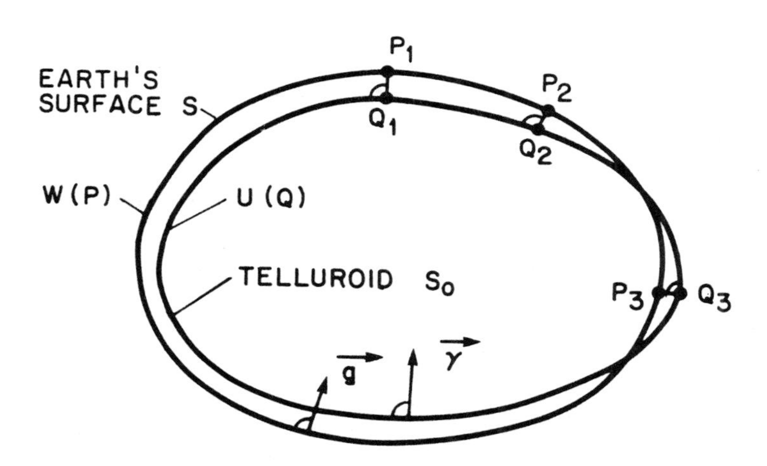

Figure 1: Definition of the reference surface S_o (telluroid)

Molodensky then simplified the problem further by introducing spherical approximation and solved, what is now known as the simple Molodensky problem, by representing T as the potential of a surface layer and transforming the boundary condition into a singular integral equation. The solution is obtained by a series expansion, the first term of which is Stokes' integral. Molodensky's solution is still widely used for numerical computations because of its intuitive appeal as a refinement of Stokes' solution.

In the early seventies, renewed interest in boundary value problems led to a rigorous linearization of the problem by Meissl (1971) and Krarup (1973) from which much of the subsequent work originated. The linearized problem can be formulated (Sanso, 1981): Find the anomalous potential T and the constant vector $\vec{a} = (a_1, a_2, a_3)$ such that

$$\Delta T = 0 \quad \text{outside } S_o$$

$$T - \gamma \left(\frac{\partial \gamma}{\partial x} \right)^{-1} \text{grad } T = \Delta g + \sum_i a_i A_i \quad \text{on } S_o \tag{18}$$

$$T(r) = \frac{c}{r} + O(r^{-3}) \quad \text{as } r \to \infty$$

where γ is the reference gravity on S_o, x is normal to S_o, $(\partial\gamma/\partial x)$ is regular in a region near S_o, $\vec{a}$ is a vector of three unknown constants, A_i are the first-order harmonics on S_o, and c is a constant. The formulation shows the invariance of the data set under arbitrary translations of the coordinate system which is a characteristic of Molodensky's problem. General solutions of this problem have not been achieved but interesting partial results are available.

The first comprehensive analysis of the linearized problem was given by Hörmander (1976) who used an implicit function theorem for the solution. He derived fundamental existence and uniqueness theorems for a boundary surface of sufficient smoothness. A study whether or not the Earth's surface satisfies the conditions of Hörmander's solution, is still missing. However, the theorem seems to be general enough to ensure the uniqueness of the solution for the actual topography and an ellipsoidal reference gravity field. The existence of the solution can, in general, be obtained from the regularity of the oblique-derivative problem for which the Fredholm alternative holds. Somewhat stronger results for the linear problem have been obtained by Sanso (1981). Hörmander also treated the nonlinear problem in the same paper and asserted existence of the solution for all W and g in a $H^{2+\varepsilon}$ neighbourhood of W_o and g_o and $\vec{a}$ close to zero in R^3. H^α is the space of k-times differentiable continuous functions whose k-th derivatives satisfy a Hölder condition with $\alpha - k \leq 1$. Uniqueness of the solution is ensured for a $H^{3+\varepsilon}$ neighbourhood. Although of considerable value from a theoretical point of view, it appears that the smoothness conditions of Hörmander's theorem cannot be met if one considers the actual topography of the Earth.

At about the same time, Sanso (1977) published a solution of the nonlinear problem based on the Legendre transformation. This method, when applied to the geodetic boundary value problem, has become known as the gravity space approach. The boundary operator has an especially simple form in 'gravity coordinates' and the transformation into a fixed boundary problem is direct, thus eliminating the use of perturbation techniques. Results achieved so far assert existence and uniqueness of the solution in a $H^{1+\varepsilon}$ neighbourhood, see Sanso (1981). In addition to these less restrictive conditions, the solution has the advantage that it does not directly depend on the terrain inclination as Hörmander's solution does, thus eliminating the use of a smoothed topography.

Recent activity in the field of boundary value problems has concentrated on the mixed boundary value problem or the so-called altimetry-gravimetry problem. Making use of the fact that gravity data are mainly available on land while altimeter data are only given for ocean areas, a boundary value problem for this mutually exclusive data situation is formulated. On land (S_1) we have again a free boundary value problem of oblique derivative type satisfying the Laplace equation. On the oceans (S_2), we have a Dirichlet problem because S_2 is known from altimeter measurements and can in first approximation be considered as equal to the geoid W_2=constant. A discussion of the resulting boundary value problems and their linearization is given by Sanso (1983). When neglecting the small difference between mean sea surface and potential, the so-called sea surface topography, the linearized problem can be formulated (Sanso, 1981): Find T regular at infinity such that

$$\Delta T = 0 \quad \text{outside } S_o$$

$$T - \left(\frac{\partial \gamma}{\partial x} \right)^{-1} \gamma \frac{\partial T}{\partial x} = \Delta W - \left(\frac{\partial \gamma}{\partial x} \right)^{-1} \gamma \, \Delta g \quad \text{on } S_o^1 \tag{19}$$

$$T = h \quad \text{on } S_o^2$$

where S_o is again an approximating surface composed of S_o^1 on land and S_o^2 on the oceans, $\Delta W = W(P) - U(Q)$, h is the ellipsoidal height, and all other terms have been explained previously. The formulation of the nonlinear problem and of a slightly different linearized problem is due to Holota (1980). He also derived the first uniqueness and existence results for the spherical case using variational methods. Although these first results were somewhat restrictive with respect to the size of the continents - a maximum spherical cap diameter of about 5° was allowed - they opened the way to further studies by the same author and by Sacerdote and Sanso (1983). The most general results so far have been obtained by Svensson (1983) who used the theory of pseudodifferential operators to obtain existence and uniqueness for data and solution belonging to a suitable Sobolev space. Wellposedness of the problem has been proved for continental areas smaller than about 63°

in diameter. The paper proposes the use of finite element methods for numerical computations and estimates convergence rates.

Recently, Sacerdote and Sanso (1985) have discussed a class of overdetermined boundary value problems which will be of considerable interest to geodesy if present plans for new data acquisition systems become reality. They will provide new data in areas where data sets already exist. Thus, the problem of dealing with heterogeneous data sets on the same boundary surface has to be treated. The data are in general inconsistent due to measurement errors. The proposed solution takes the stochastic nature of these errors into account and achieves boundedness of the estimator by modifying the minimum principle by a regularization technique. There is obviously a close connection between these problems and the problems to be treated in the next chapter.

5. APPROXIMATION OF T IN A REPRODUCING KERNEL HILBERT SPACE AND REGULARIZATION

In this chapter, approaches to deal more realistically with the actual data situation, will be discussed. Thus, heterogeneous data sets will be considered which are not necessarily given at the surface of the Earth but may be at various altitudes above it. In addition, a regular data distribution or a complete global coverage will not be assumed. Since a comprehensive discussion of current techniques and typical geodetic applications is contained in Moritz (1980), the presentation will again be restricted to outlining the results available and the remaining open problems.

The approximation of T in a reproducing kernel Hilbert space was proposed in Krarup (1969). His paper was seminal in outlining the mathematical problem behind some statistical estimation techniques which at that time were used rather indiscriminately in geodesy. It established a foundation for an approximation of T from discrete, heterogeneous data and directed the discussion towards the underlying mathematical problems. Central to his approach is the proof of Runge's theorem for the geodetic case. It is given as

> Given any potential regular outside the surface of the earth and any sphere in the interior of the earth. For every closed surface surrounding the earth (which surface may be arbitrarily near the surface of the earth) there exists a sequence of potentials regular in the whole space outside the given sphere and uniformly converging to the given potential on and outside the given surface.

The consideration of such a sequence by least-squares methods is discussed in the paper.

The problem as proposed by Krarup can be formulated in the following way (Sanso and Tscherning, 1980): Let the anomalous potential T be a function harmonic outside the surface S of the Earth and let measurements δb_j, which have been performed on linear functionals L_j of T, be available on S and outside it. Find an approximation $\hat{T}$ to T such that

(a) $\hat{T}$ is harmonic outside a sphere S_o which is totally enclosed in the Earth. The set in R^3 outside S_o will be called Ω.

(b) $\hat{T}$ is a member of a Hilbert space $H_K(\Omega)$ of harmonic functions with reproducing kernel K(P,Q).

(c) $\hat{T}$ represents the same measurements δb_j as T, i.e.

$$L_j \hat{T}(P) = L_j T(P) = \delta b_j$$

where L_j are bounded evaluation functionals.

(d) $\hat{T}$ satisfies

$$|| \hat{T} || = \text{minimum}.$$

The solution of the above problem as given by Krarup is a collocation solution of the form

$$\hat{T}(P) = \{L_i K(P,Q)\}^T \{L_j L_i K(P,Q)\}^{-1} \delta b_j \tag{20}$$

where $\{\cdot\}$ has been used to denote matrices and K(P,Q) is defined by (15). For $T \in H_K$, strong convergence of $\hat{T}$ towards T in H_K can be proved iff the sequence $L_i K(P,Q)$ is complete in H_K. The case $T \notin H_K$ has been discussed in Krarup (1978) for a hybrid norm.

A related convergence problem has been treated in Moritz (1976) where the set of harmonicity is the set outside the Earth's surface. In this case, arbitrarily good approximations can be found if the chosen Hilbert space has T as an element and if the linear functionals associated with δb_j form a complete set in the dual space.

In practical applications of collocation, an empirically determined covariance function C(P,Q) is often used as the reproducing kernel. It is obtained by requiring C(P,Q) to be invariant under rotations on the sphere. In this case, T is harmonic in Ω outside the sphere and is not an element of the Hilbert space H_C with the reproducing kernel C(P,Q). The question then arises whether or not replacing K(P,Q) by C(P,Q) in equation (20) will result in an approximation $\hat{T}$ that converges towards T. This problem is discussed in Sanso and Tscherning (1980). They show that, in general, convergence cannot be proved. However, if the data and T are smoothed in the same manner, convergence can be shown to exist. This case is of practical importance because data smoothing is a common preprocessing step. More recently, Barzaghi and Sanso (1985) have shown asymptotical convergence of the smoothed sequences $\hat{T}_s$ towards T. It appears, however, that in light of Svensson's (1983) recent results, the discussion on the validity of this method for the mixed data problem is far from finished.

The problem of downward continuing data from flight and satellite altitudes to the Earth's surface was early recognized as an improperly posed problem. Moritz (1966) analyzed downward continuation of airborne gravity measurements in this context and discussed some classical solution approaches. Schwarz (1971) treated the same problem using Tikhonov's regularization technique to minimize the nonlinear functional

$$M^{\alpha}\{g, \delta b\} = ||LT - \delta b|| + \alpha\, \Omega(g) \tag{21}$$

where α is a numerical parameter and $\Omega(g)$ is the stabilizing functional. Using

$$\Omega(g) = ||\, g\, ||$$

led to a modified collocation solution. The relation of the two methods was further discussed in Schwarz (1979), where some simulation results are given, and in Moritz (1980). A presentation of geodetic improperly posed problems in the framework of inverse potential theory was first given by Grafarend (1972), and more recently Anger, in a series of papers, has considerably extended the discussion; for a review and additional literature see Anger (1981). A systematic treatment of geodetic problems using regularization techniques has been provided by the Russian school. Key papers are Neyman (1976), Tikhonov et al. (1978), and Neyman (1984). Especially the last paper gives a detailed treatment of many relevant questions, discusses solutions for geodetic problems of the integral equation and derivative type, and deals with some implementation aspects as e.g. the non-iterative determination of a suboptimal α. Despite these efforts, regularization methods have not been implemented to any large extent into standard geodetic computations. The discussion of possible reasons for this situation will therefore conclude this chapter.

There is at present, apparently no well-developed theory to simultaneously use heterogeneous data types, as e.g. first and second order gradients of T, in the regularization technique. Thus, a standard data situation in geodesy is not covered by the methods available. Simulation studies show that the use of different data types tends to make solutions more stable. Figures 2 and 3, taken from Schwarz (1979), illustrate this point. They show regularized solutions for the downward continuation of airborne gravity, using in the first case, gravity data only, and in the

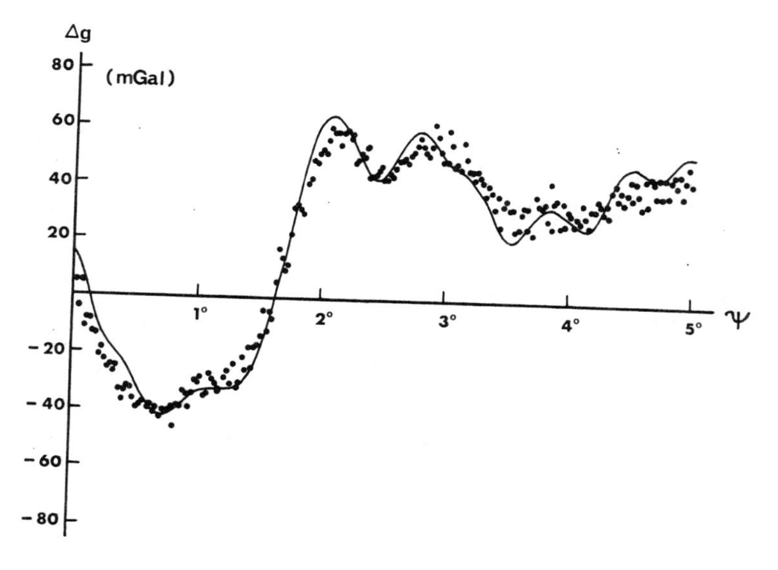

Figure 2: Downward continuation of gravity from aircraft altitudes using a regularized solution for gravity only.

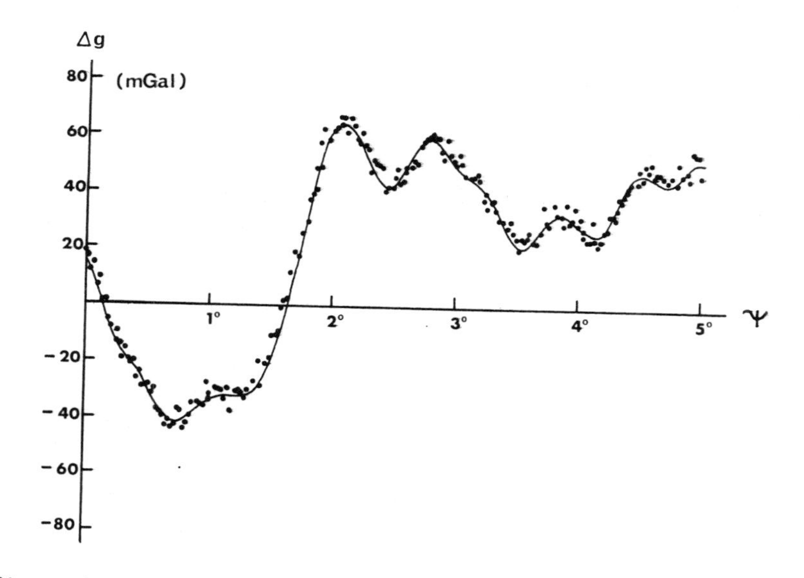

Figure 3: Downward continuation of gravity from aircraft altitudes using a regularized solution for gravity only.

second case, gravity data and two gravity gradients. The estimates are given as points, the known solution as a solid line. For the first data set, the approximation is clearly biased, despite an optimal α, while this is not the case for the second data set. A regularization approach using observations of T and its derivatives simultaneously would certainly have wide application in geodesy. The second impediment to the wider use of regularization techniques lies in the fact that the data set used and the coefficient set to be determined are extremely large. A typical situation, seen to be treated in airborne gravity gradiometry, is a data set of 250 000 values and a coefficient set of at least 50 000. Thus, iterative methods to determine an optimal α or cross-validation techniques are often too time consuming to be applied. Methods to determine a priori a suboptimal α 'close' to the optimal value, would therefore be of considerable help. Similarly, strategies to subdivide a given data set and to improve the solution step by step, would alleviate current problems to some extent. Thus, the use of regularization techniques in geodesy is closely linked to the question whether these methods can be implemented for large data and coefficient sets.

REFERENCES

Anger, G., "A characterization of the inverse gravimetric source problem through extremal measures", Review of Geophys. and Space Phys., 19 (1981), 299-306.

Barzaghi, R. and F. Sanso, "New results on the convergence problem in collocation theory", First Hotine-Marussi Symposium on Mathematical Geodesy, Rome, June 3-6, (1985).

Freeden, W., "On the permanence property in spherical spline interpolation", Rep. 341, Dep. of Geod. Sci., Ohio State University, Columbus, (1982).

Grafarend, E.W., "Inverse potential problems of improperly posed type in physical geodesy and geophysics", Symp. on Earth Gravity Models and Related Problems, St. Louis, (1972).

Grafarend, E.W., B. Heck, E.H. Knickmeyer, "The free versus fixed geodetic boundary value problem for different combinations of geodetic observables", Bull. Géod., 59 (1985), 11-32.

Holota, P., "The altimetry-gravimetry boundary value problem", Proc. Int. Conf. Intercosmos, Albena, (1980).

Hörmander, L., "The boundary value problem of physical geodesy", Rep. 9, Mittag Leffler Inst., Stockholm, (1975).

Klein, F., Elementarmathematik vom höheren Standpunkt III, 3rd edition, Springer Verlag, Berlin, (1928).

Koch, K.R., "Die geodätische Randwertaufgabe bei bekannter Erdoberfläche", Zeitschr. für Verm., 96 (1971), 218-224.

Krarup, T., "A contribution to the mathematical foundation of physical geodesy", Geodaetisk Instituts Meddelelse No. 44, Kobenhavn, (1969).

Krarup, T., "A convergence problem in collocation theory", 7th Symposium on Mathematical Geodesy, Assisi, (1978).

Meissl, P., "On the linearization of the geodetic boundary value problem", Rep. 152, Dep. of Geod. Sci., Ohio State University, Columbus, (1971).

Meissl, P., "The use of finite elements in physical geodesy", Rep. 313, Dep. of Geod. Sci., Ohio State University, Columbus, (1981).

Moritz, H., "Methods for downward continuation of gravity", DGK, Series A, No. 30, Munich, (1966).

Moritz, H., "Integral formulas and collocation", Manuscripta Geodaetica, 1, (1976), 1-40.

Moritz, H., Advanced physical geodesy, H. Wichmann Verlag, Karlsruhe, (1980).

Neyman, Yu. M., "Justification of the variational method of the theory of the shape of the earth", Geod., Map. and Photogram., 18 (1976), 216-218.

Neyman, Yu. M., "Improperly posed problems in geodesy and methods of their solution", in Schwarz (ed.) Proc. Local Gravity Field Approximation, Beijing, Aug. 21-Sept. 4, 1984, UCSE Rep. 60003, University of Calgary, (1985).

Rapp, R.H. "The earth's gravity field to degree and order 180 using Seasat altimeter data, terrestrial gravity data and other data", Rep. 322, Dep. of Geod. Sci., Ohio State University, Columbus, (1981).

Rummel, R., "The determination of gravity anomalies from geoid heights using the inverse Stokes formula", Rep. 269, Dep. of Geod. Sci., Ohio State University, Columbus, (1977).

Sacerdote, F. and F. Sanso, "A contribution to the analysis of the altimetry-gravimetry problem", in Holota (ed.) Proc. Figure of the Earth, Moon and other Planets, Prague, Sept. 20-25, (1982).

Sacerdote, F. and F. Sanso, "Overdetermined boundary value problems in physical geodesy", Manuscripta Geodaetica, 10 (1985), 195-207.

Sanso, F., The geodetic boundary value problem in gravity space, Atti R. Accad. Naz. Lincei Mem. Cl. Sci. Fis. Mat. Nat., 16, (1977) 1-52.

Sanso, F., "Recent advances in the theory of the geodetic boundary value problem", Rev. of Geophys. and Space Phys., 19 (1981) 437-449.

Sanso, F., "A discussion on the altimetry-gravimetry problems", in K.P. Schwarz and G. Lachapelle (eds.) Geodesy in Transition, UCSE Rep. 60002, University of Calgary, (1983).

Sanso, F. and C.C. Tscherning, "Notes on convergence in collocation theory", Boll. di Geod. e. Sci. Affini, 39 (1980), 221-252.

Schwarz, K.P., "Numerische Untersuchungen zur Schwerefortsetzung", DGK, Series C, No. 171, Munich, (1971).

Schwarz, K.P., "Geodetic improperly posed problems and their regularization", Boll. di Geod. e. Sci. Affini, 38 (1979), 389-416.

Schwarz, K.P., "Data types and their spectral properties", in Schwarz (ed.) Proc. Local Gravity Field Approximation, Beijing, Aug. 21-Sept. 4, 1984, UCSE Rep. 60003, University of Calgary, (1985).

Svensson, S.L., "Solution of the altimetry-gravimetry problem", Bull. Géod., 57 (1983), 332-350.

Tikhonov, A.N., V.D. Bol'Shakov, Yu.M. Neyman, "Incorrect problems in geodesy", Geod., Map. and Photogram., 20 (1978), 146-151.

Tscherning, C.C. "A note on the choice of norm when using collocation for the computation of approximations to the anomalous potential", Bull. Géod., 51, (1977), 137-147.

Tscherning, C.C. and R.H. Rapp, "Closed covariance expressions for gravity anomalies, geoid undulations, and deflections of the vertical implied by anomaly degree variance models", Rep. 208, Dep. of Geod. Sci., Ohio State University, (1974).

Tscherning, C.C., R.H. Rapp, C. Goad, "A comparison of methods for computing gravimetric quantities from high degree spherical harmonic expansions", Manuscripta Geodaetica, 8 (1983), 249-272.

DIVISION OF SURVEYING ENGINEERING
THE UNIVERSITY OF CALGARY
2500 UNIVERSITY DRIVE N.W.
CALGARY, ALBERTA, CANADA
T2L 1N7

Contemporary Mathematics
Volume **59**, 1986

WHAT REGRESSION MODEL SHOULD BE CHOSEN WHEN THE STATISTICIAN MISSPECIFIES THE ERROR DISTRIBUTION ?

Wolfgang Härdle[1]

ABSTRACT. We consider the situation where the statistician fails to chose the correct Likelihood function in a regression model. We propose a model selection rule and show its asymptotic optimality. Relationships to C_p and extensions of AIC are discussed.

1. INTRODUCTION. Let $\underset{\sim}{Y}_n=(Y_1,\dots,Y_n)'$ be a random vector of n independent observations with mean vector $\underset{\sim}{\mu}_n=(\mu_1,\dots,\mu_n)'$. Assume that the ith mean μ_i is associated with an infinite covariate x_i in a linear way, i.e.

$$\mu_i = \langle x_i,\beta\rangle$$

where the parameter β and the covariate are in ℓ_2, the space of square summable sequences equipped with the canonical inner product.Suppose that the observation error $e_i=Y_i-\mu_i$ has distribution F with density f. In general the statistician does not know f, so he might fix a different error density h and use the model

$$H_p=\{\prod_{i=1}^{n} h(Y_i-\langle x_i,\beta(p)\rangle):\beta'(p)=(0,\dots,\beta_{p_1},0,\dots,\beta_{p_2},\dots,\beta_{p_{k(p)}},0,\dots)\}$$

where $p_1<p_2<\dots<p_{k(p)}$ and $k(p)\geq 1$ is the dimension. Such a mismatch of the chosen model and the true error distribution can happen in a variety of cases. For instance, the statistician could apply a robust regression procedure (Huber, 1973) but the data is in fact Gaussian. We may also imagine the reverse situation. A natural way of evaluating goodness of the regression model H_p is to introduce some kind of distance between the predicted re-

[1] Partially supported by Deutsche Forschungsgemeinschaft SFB 123 "Stochastische Mathematische Modelle".

ression surfaces and the true model regression $\underset{\sim}{\mu}_n$. We consider here the Euklidean distance

$$L_n(p) = \|\underset{\sim}{\mu}_n - \hat{\underset{\sim}{\mu}}_n(p)\|^2,$$

where $\hat{\underset{\sim}{\mu}}_n(p)=\langle x_i,\hat{\beta}(p)\rangle$ denotes the predicted regression surface based on the maximum likelihood estimate $\hat{\beta}(p)$ in model H_p. The assumed error density h is thought of beeing fixed, so the above loss $L_n(p)$ depends only on

$$p = (p_1, p_2, \ldots, p_{k(p)})$$

which we call from now on *the model* p.

Which model p should be selected if a variety P_n of models is possible? In this paper we derive an efficient model selection procedure and prove that it asymptotically minimizes $L_n(p)$ over a set of models P_n. Results of this type have been obtained by Shibata (1981), Breiman and Friedman (1983) in the case of $h \equiv f \equiv \varphi$, the density of the normal distribution function. Recently Li (1984) gave conditions for asymptotic efficiency of least squares estimators using model choice procedures based on cross validation, FPE among others. In these special cases our procedure is equivalent. It is also related to an extension of AIC given by Takeuchi (1976). This connection is investigated in Section 3, the main result in the next section.

2. ASYMPTOTIC EFFINCIENCY WHEN THE REGRESSION MODEL IS MISSPECIFIFIED. Define the linear operator $X_n: \ell_2 \to \mathbb{R}^n$ by $X_n'=(x_1'x_2'\ldots x_n')$, the vector of observation errors $\underset{\sim}{e}_n=(e_1,e_2,\ldots,e_n)'$ and $\psi(u)=-\frac{d}{du}\log h(u)$, $\gamma = E_F\psi^2(e)/(E_F\psi'(e))^2$. An expansion of $L_n(p)$ motivates the score

$$W_n'(p) = -\|\hat{\underset{\sim}{\mu}}_n(p)\|^2 + 2\gamma k(p) + \|\underset{\sim}{\mu}_n\|^2 .$$

To see this, observe that

$$\begin{aligned} W_n'(p)-L_n(p) &= -\{\|\hat{\underset{\sim}{\mu}}_n(p)-\underset{\sim}{\mu}_n\|^2 + 2\langle\hat{\beta}(p)-\beta,\beta\rangle_{B_n} + \|\underset{\sim}{\mu}_n\|^2\} \\ &\quad + 2\gamma k(p) + \|\underset{\sim}{\mu}_n\|^2 - \|\hat{\underset{\sim}{\mu}}_n(p) - \underset{\sim}{\mu}_n\|^2 \\ &= -2\|\hat{\underset{\sim}{\mu}}_n(p)-\underset{\sim}{\mu}_n\|^2 + 2\gamma k(p) \qquad (1) \\ &\quad -2\{-\|\hat{\underset{\sim}{\mu}}_n(p)-\underset{\sim}{\mu}_n\|^2 + \langle\hat{\beta}(p)-\beta,\hat{\beta}(p)\rangle_{B_n}\} \\ &= 2\{\gamma k(p) - \langle\hat{\beta}(p)-\beta,\hat{\beta}(p)\rangle_{B_n}\}, \end{aligned}$$

where $\langle u,v\rangle_{B_n} = u'B_n v$, $B_n = X_n'X_n$.

Suppose that the last term in (1) is tending to a constant uniformly over models $p \in P_n$. Then minimizing $W_n'(p)$ over P_n will be the same task, at least asymptotically, as minimizing $L_n(p)$. Two unknowns are still involved in $W_n'(p)$. The constants $\|\mu_n\|^2$ and γ depend on the unknown regression function and the unknown true error distribution F. The constant $\|\mu_n\|^2$ does not cause difficulties since it is independent of the model: So minimizing $W_n'(p) - \|\mu_n\|^2$ is the same as minimizing $W_n'(p)$. The scaling factor γ can be estimated by a consistent sequence of estimators

$$\hat{\gamma}_n = \frac{n^{-1} \sum_{i=1}^{n} \psi^2(Y_i - \langle x_i, \hat{\beta}(p_n)\rangle)}{[n^{-1} \sum_{i=1}^{n} \psi'(Y_i - \langle x_i, \hat{\beta}(p_n)\rangle)]^2}$$

where $\{p_n\}$ is a model sequence of increasing dimension.
We will therefore define

$$W_n(p) = -\|\hat{\mu}_n\|^2 + 2\hat{\gamma}_n k(p)$$

as the score function that is to be minimized over P_n. The concept of *asymptotic efficiency* is defined as follows.

DEFINITION. A selected $\hat{p}$ is called asymptotically efficient if, as $n \to \infty$,

$$\frac{L_n(\hat{p})}{\inf_{p \in P_n} L_n(p)} \xrightarrow{P} 1 .$$

The following conditions are needed.

CONDITION 1. The function ψ is twice differentiable and fulfills

(i) $E_F \psi(e) = 0$

(ii) $q = E_F \psi'(e) > 0$

(iii) $\exists N > 0: \quad E_F[q^{-1}(\psi'(e) - q)]^{2N} < \infty$.

CONDITION 2. The matrix $B_n(p) = X_n'(p) X_n(p)$ has full rank $k(p)$, where $X_n(p)$ is the (n,p) matrix containing only the non-zero control variables of model p. There exists a $N>0$ such that

$$\sum_{p\in P_n} \tilde{R}_n(p)^{-N} \to 0 \ , \quad \text{as} \quad n \to \infty,$$

where $\tilde{R}_n(p) = E_F\tilde{L}_n(p)$, $\tilde{L}_n(p) = ||\mu_n - \tilde{\mu}_n(p)||^2$ and $\tilde{\mu}_n(p)$ denotes the Gauß-Markov estimator $X_n(p)B_n^{-1}(p)X_n'(p)\tilde{Y}_n$. applied to the *pseudodata* $Y_i = \mu_i + \tilde{e}_i$, $\tilde{e}_i = \psi(e_i)/q$.

CONDITION 3. Let $h(p)$ be the largest diagonalelement of the *hat matrix* $M_n(p) = X_n(p)B_n^{-1}(p)X_n'(p)$.

Assume $\sup_{p\in P_n} h(p)\tilde{R}_n(p) \to 0$, as $n \to \infty$.

THEOREM. *Choose* $\hat{p}$ *such that it minimizes* $W_n(p)$ *over* P_n. *Under conditions* 1-3, $\hat{p}$ *is asymptotically efficient.*

REMARK 1. The estimates $\hat{\beta}(p)$ will be compared with the Gauß-Markov estimates in model p, that are based on the (non observable) pseudodata $\tilde{Y}_i = \mu_i + \tilde{e}_i$. It will then be seen that the problem of asymptotic efficiency can be solved by an analogeous problem formulated for the linear estimates $\tilde{\mu}_n(p)$. Details of the proof of the theorem are contained in Härdle (1985).

REMARK 2. It follows from condition 2 that $k^2(p)/n \to 0$, as $n \to \infty$. This is an analogue of the (necessary) condition "$p^2/n \to 0$", that can be found in Huber (1981, p.166).

REMARK 3. If ψ is a bounded function, as is assumed in robust regression analysis, condition 2 can be weakened. It is seen from the proof in Härdle (1985) that $\sum_{p\in P_n} \exp(-c\tilde{R}_n(p)) \to 0$, $c > 0$ is sufficient.

3. CONNECTION TO OTHER METHODS. In the case of least squares estimators there are a variety of model selection procedures, such as generalized gross validation, FPE, AIC or Mallows' (1973) C_p. Shibata (1981) and Li (1984) have shown the equivalence of these procedures. The linearization of $W_n(p)$ i.e. the score function W_n based on the nonobservable pseudodata $\tilde{Y}_n$ has a similar structure as C_p. The latter score for $\tilde{Y}_n$ reads

$$C_p = \|\tilde{\underset{\sim}{Y}}_n - \tilde{\underset{\sim}{\mu}}_n\|^2 + 2\gamma k(p)$$
$$= \|\tilde{\underset{\sim}{e}}_n\|^2 + \tilde{L}_n(p) + 2\tilde{\underset{\sim}{e}}'(I_n - M_n(p))\underset{\sim}{\mu}_n$$
$$+ 2\{\gamma k(p) - \tilde{\underset{\sim}{e}}' M_n(p)\tilde{\underset{\sim}{e}}\} .$$

Expand $W_n(p)$ with $\hat{\gamma}_n$ replaced by γ

$$W_n(p) = -\|\tilde{\underset{\sim}{\mu}}_n(p)\|^2 + 2\gamma k(p)$$
$$= -\|\tilde{\underset{\sim}{\mu}}_n(p) - \underset{\sim}{\mu}_n\|^2 - \|\underset{\sim}{\mu}_n\|^2 - 2(\tilde{\underset{\sim}{\mu}}_n(p) - \underset{\sim}{\mu}_n)'(\underset{\sim}{\mu}_n - \tilde{\underset{\sim}{\mu}}_n(p))$$
$$- 2(\tilde{\underset{\sim}{\mu}}_n(p) - \underset{\sim}{\mu}_n)'\tilde{\underset{\sim}{\mu}}_n(p) + 2\gamma k(p)$$
$$= \tilde{L}_n(p) + 2\gamma k(p) - 2\tilde{\underset{\sim}{e}}_n' M_n(p)\tilde{\underset{\sim}{e}}_n + 2\tilde{\underset{\sim}{e}}_n'(I_n - M_n(p))\underset{\sim}{\mu}_n$$
$$+ 2\underset{\sim}{\mu}_n'\tilde{\underset{\sim}{e}}_n - \|\underset{\sim}{\mu}_n\|^2.$$

The last two terms are independent of p. The remaining terms are identical to those in Mallows' C_p.

There is a different way of deriving $W_n(p)$. Our way was to argue that $\hat{\underset{\sim}{\mu}}_n(p)$ is asymptotically like the least square estimator $\tilde{\underset{\sim}{\mu}}_n(p)$ based on the unobservable pseudodata $\tilde{\underset{\sim}{Y}}_n$. This made it possible to argue as in the linear case. Takeuchi (1976) argued in a different way, when he heuristically extended Akaike's (1970) AIC to the case of mismatching the true likelihoodfunction. Takeuchi gave no proof, but his derivation is interesting, we therefore want to present it here again.

Denote by $I(f, h_{\beta(p)})$ the Kullback-Leibler information number between f and $h_{\beta(p)} \in H_p$. Consider the prediction error $E^Y(I(f, h_{\hat{\hat{\beta}}(p)}))$, where $\hat{\hat{\beta}}(p)$ is the maximum likelikood estimator, which is based on $\{(x_i, Y_n')\}_{i=1}^n$ a data set with $\{Y_i'\}$ distributed as $\{Y_i\}$. Write

$$E^Y(I(f, h_{\hat{\hat{\beta}}(p)})) = E^Y \int f(u) \log \frac{f(u)}{h(u;\hat{\hat{\beta}}(p))} du$$
$$= \int f(u) \log f(u) du - E^Y \int f(u) \log h(u;\hat{\hat{\beta}}(p)) du$$

and observe that the first term is independent of the model p. Expand

$$\log h(u,\hat{\hat{\beta}}(p)) = \log h(u;\beta^*(p)) + (\hat{\hat{\beta}}_{(p)} - \beta_{(p)})\frac{\partial}{\partial\beta} \log h(u;\beta^*_{(p)})$$
$$+ \frac{1}{2}(\hat{\hat{\beta}}(p) - \beta^*(p))'(\frac{\partial^2}{\partial\beta\partial\beta'} \log h(u;\beta^*(p)))(\hat{\hat{\beta}}(p) - \beta^*(p))$$
$$+ \ldots ,$$

where $\beta^*(p)$ minimizes $I(f, h_{\beta^*(p)})$.
Then

$$\int f(u) \log f(u;\hat{\hat{\beta}}(p))du$$

$$\doteqdot \int f(u) \log f(u;\beta^*(p))du - \tfrac{1}{2}(\hat{\hat{\beta}}(p)-\beta \;\; (p)'J(\hat{\hat{\beta}}(p)-\beta^*(p)$$

where

$$J = E(-\frac{\partial^2}{\partial\beta\partial\beta'} \log h(u;\beta^*)).$$

Since the maximum likelihood estimator is asymptotically normally distributed

$$\sqrt{n}\,(\hat{\hat{\beta}}(p)-\beta^*(p)) \xrightarrow{L} N(0,\ J^{-1} I J^{-1})$$

with $I = E(\frac{\partial}{\partial\beta} \log h(u;\beta^*(p))\frac{\partial}{\partial\beta'} \log h(u;\beta^*(p))$

it follows that

$$E^Y \int f(u) \log h(u;\hat{\hat{\beta}}(p))du \doteqdot \int f(u) \log h(u;\beta^*(p))du - tr(IJ^{-1}) / 2n.$$

On the other hand for the data set $\{(X_i, Y_i)\}_{i=1}^n$

$$\begin{aligned}\log h(\underset{\sim}{Y}_n,\beta^*(p)) &= \log h(\underset{\sim}{Y}_n,\hat{\beta}(p)) \\ &+ (\beta^*(p)-\hat{\beta}(p))'\frac{\partial}{\partial\theta} \log h(\underset{\sim}{Y}'_n,\hat{\beta}(p)) \\ &+ \tfrac{1}{2}(\beta^*(p)-\hat{\beta}(p))'\frac{\partial}{\partial\theta\partial\theta'} \log h(\underset{\sim}{Y}_n, \hat{\beta}(p))(\beta^*(p)-\hat{\beta}(p))\end{aligned}$$

where $\hat{\beta}(p)$ is the maximum likelihood estimator based on $\{(X_i, Y_i)\}_{i=1}^n$.

Then

$$\int f(u) \log h(u;\beta^*(p))du = n^{-1}\int f(\underset{\sim}{Y}_n - X_n\beta) \log h(\underset{\sim}{Y}_n;\beta^*(p))d\underset{\sim}{Y}_n$$

$$\doteqdot n^{-1}\int f(\underset{\sim}{Y}_n - X_n\beta) \log h(\underset{\sim}{Y}_n;\hat{\beta}(p))d\underset{\sim}{Y}_n - tr(IJ^{-1})/2n\ ,$$

and

$$E^Y \int f(u) \log h(u;\hat{\hat{\beta}}(p))du$$

$$\doteqdot n^{-1}\int f(\underset{\sim}{Y}_n - X_n\beta) \log h(Y_n;\hat{\hat{\beta}}(p))d\underset{\sim}{Y}_n - \tfrac{1}{n}tr(IJ^{-1}).$$

Concequently the task to minimize $E^Y(I(f;h_{\hat{\beta}(p)}))$ is the same as to minimize the approximate quantity

$$-E(\log h(\underset{\sim}{Y}_n;\hat{\beta}(p))+tr(IJ^{-1})$$

and if we replace the expectation by the observation that we have at hand we obtain

$$GAIC(p) = -2 \log h(\underset{\sim}{Y}_n;\hat{\beta}(p)) + 2tr(IJ^{-1}).$$

which we call *Generalized AIC*.

In the regression setting we have

$$\log h(\underset{\sim}{Y}_n;\beta(p)) = \sum_i \log h(Y_i - \langle x_i, \beta(p)\rangle)$$

$$IJ^{-1} = X' \begin{pmatrix} V_1 & \cdots & 0 \\ \vdots & \ddots & \\ 0 & \cdots & V_n \end{pmatrix} X \; X' \begin{pmatrix} W_1 & \cdots & 0 \\ \vdots & \ddots & \\ 0 & \cdots & W_n \end{pmatrix} X^{-1}$$

where $V_i = \mathrm{var}(\psi(Y_i - \langle x_i, \beta^*(p)\rangle)$

$W_i = E\psi'(Y_i - \langle x_i, \beta^*(p)\rangle)$.

A Taylor expansion shows that

$$V_i \doteq E\psi^2(e_i) \quad \text{if} \quad \langle x_i, \beta - \beta^*(p)\rangle \to 0$$

$$W_i \doteq E\psi'(e_i)$$

Therefore $\mathrm{tr}(IJ^{-1}) \doteq p \cdot \frac{E\psi^2}{E\psi'}$.

Now

$$\log h(\underset{\sim}{Y}_n;\beta(p)) = \log h(\underset{\sim}{Y}_n;\hat\beta(p)) + (\beta-\hat\beta(p))\frac{\partial}{\partial\beta}\log h(\underset{\sim}{Y}_n,\hat\beta(p))$$

$$+ \frac{1}{2}(\beta-\hat\beta(p))' \frac{\partial^2}{\partial\beta\partial\beta} \log h(\beta-\hat\beta(p)) + \ldots$$

$$\doteq \log h(\underset{\sim}{Y}_n;\hat\beta(p)) - \frac{1}{2}(\beta-\hat\beta(p))'J(\beta-\hat\beta(p)).$$

Therefore

$$\log h(\underset{\sim}{Y}_n;\hat\beta(p)) = \log h(\underset{\sim}{Y}_n;\beta) + \frac{1}{2}\|\beta-\hat\beta(p)\|_J^2$$

$$= \log h(\underset{\sim}{Y}_n;\beta) + \frac{1}{2}\|\beta\|_J^2 - \langle\beta,\hat\beta(p)\rangle_J + \frac{1}{2}\|\hat\beta(p)\|_J^2$$

and

$$-2\log h(\underset{\sim}{Y}_n;\hat\beta(p)) \doteq -2\log h(\underset{\sim}{Y}_n;\beta) - \|\beta\|_J^2 + 2\langle\beta,\hat\beta(p)\rangle_J - \|\hat\beta(p)\|_J^2$$

$$= -2\log h(\underset{\sim}{Y}_n;\beta) - \|\beta\|_J^2 + 2\langle\beta,\hat\beta(p)-\beta\rangle_J + 2\|\beta\|_J^2$$

$$- \|\hat\beta(p)\|^2.$$

The crossterm tends to zero as $\langle x_i;\beta-\beta^*(p)\rangle \to 0$.
So we have that minimizing

$$\mathrm{GAIC}(p) = -2\log h(\underset{\sim}{Y}_n;\hat\beta(p)) + 2\mathrm{tr}(IJ^{-1})$$

is asymptotically the same as minimizing

$$-\|\hat\beta(p)\|_J + 2\gamma q p$$

since $J \doteq X'X/q$, this is approximately equal to $W_n(p)$.

BIBLIOGRAPHY

1. Akaike, H., "Statistical Predictor Identification", Ann. Inst. Math. Stat., 22 (1970), 203-217.

2. Breiman, L. and Freedman, D., "How many variables should be entered in a regression equation", J. Amer. Stat. Assoc., 78 (1983), 131-136.

3. Härdle, W., "An effective selection of regressive variables when the error distribution is correctly specified", submitted for publication, (1985)

4. Huber, P., "Robust regression: Asymptotics, conjectures, and Monte Carlo", Ann. Statist., 1 (1973), 799-821.

5. Huber, P., Robust Statistics, Wiley, New York, (1981).

6. Mallows, C., "Some comments on C_p", Technonetrics, 15 (1973), 661-675.

7. Li, K.C., Asymptotic optimality for C_p, C_ℓ, cross-validation and generalized cross-validation: Discrete index set., Manuscript, (1984).

8. Shibata, R., "An optimal selection of regression variables", Biometrika, 68 (1981), 45-54.

9. Takeuchi, K., "Distribution of information statistics and a criterion of model fitting", Suri Kagaku, 153 (1976), 12-18, (in Japanese).

INSTITUT FÜR GESELLSCHAFTS- UND
WIRTSCHAFTSWISSENSCHAFTEN
RHEINISCHE FRIEDRICH-WILHELMS-UNIVERSITÄT BONN
FEDERAL REPUBLIC OF GERMANY

Contemporary Mathematics
Volume 59, 1986

APPROXIMATION THEORY OF METHOD OF REGULARIZATION ESTIMATORS: APPLICATIONS

Dennis D. Cox[1]

ABSTRACT. The method of regularization is a method for estimating an unknown function from discrete noisy observations. In other works the author has developed a method for approximating the discrete method of regularization estimator which is easier to analyze. Various applications of this theory are examined here including improved rates of convergence from use of derivative data and the construction of simultaneous confidence bands.

1. INTRODUCTION. This article presents some applications of the theory in Cox(1984b, 1986a) and Cox and Nychka(1984) to ill posed linear inversion problems with discrete noisy data. We discuss an approximation of the Tikhonov(1963) method of regularization (MOR) estimator for the discrete problem by an analogous procedure applied to a "continuous" version of the problem. The main applications to be studied are: (1) the importance of derivative information; and (2) confidence regions based on the proposal of Knafl, Sacks, and Ylvisaker (1985). Other possible applications are mentioned in the closing section.

Suppose we wish to estimate an unknown function β which is an element of a Hilbert space of functions denoted Ω. (All Hilbert spaces in this paper will be real and separable, and the inner products and norms will be subscripted with the corresponding space as in $\langle\cdot,\cdot\rangle_\Omega$ and $\|\cdot\|_\Omega$.) Let X_n be a sequence of bounded linear operators mapping Ω to Hilbert spaces Y_n. The X_n are known as design operators. Suppose we have for each n a "noisy"

1980 Mathematics Subject Classification (1985 Revision). 62J99, 65D20

[1]Supported by the National Science Foundation under grant number DMS-8202560.

observation vector

$$y_n = X_n\beta + \epsilon_n , \tag{1.1}$$

where ϵ_n denotes a random vector having mean zero and covariance $\sigma_n^2 I$ w.r.t. Y_n inner product (I denotes an identity operator). This is to say that for $\forall\ \eta, \eta_1, \eta_2 \in Y_n$,

$$E\langle\epsilon_n,\eta\rangle_{Y_n} = 0 \text{ and } E[\langle\epsilon_n,\eta_1\rangle_{Y_n}\langle\eta_2,\epsilon_n\rangle_{Y_n}] = \sigma_n^2\langle\eta_1,\eta_2\rangle_{Y_n}. \tag{1.2}$$

The goal is to estimate β from y_n. The following examples will be considered in more detail below.

EXAMPLE 1 Suppose $X_n\beta = (\beta(t_{n1}), \beta(t_{n2}), \ldots, \beta(t_{nn})) \in \mathbb{R}^n$, where the t_{ni}'s are known points in [0,1]. Here, Ω is a reproducing kernel Hilbert space (so that point evaluations $\beta \mapsto \beta(t)$ are continuous) of functions with domain [0,1]. Let $Y_n = \mathbb{R}^n$ as sets. Suppose $\epsilon_n = (\epsilon_{n1}, \epsilon_{n2}, \ldots, \epsilon_{nn})$ has mean zero uncorrelated components with common variance σ^2. Then (1.2) will hold provided Y_n-inner product and σ_n^2 are given by

$$\langle\eta_1,\eta_2\rangle_{Y_n} = n^{-1}\sum_i \eta_{1i}\eta_{2i} \text{ and } \sigma_n^2 = \sigma^2/n. \tag{1.3}$$

Our reason for rescaling Y_n inner product rather than just setting $\langle\cdot,\cdot\rangle_{Y_n} = \langle\cdot,\cdot\rangle_{\mathbb{R}^n}$ will be explained in Section 2. This is the "classical" nonparametric regression setup: discrete noisy observations of a function (Nadaraya, 1964, and Watson, 1964)

One popular choice for Ω will now be described. Assume that $\beta:[0,1]\to\mathbb{R}$ and let D denote the differentiation operator. Let

$$\Omega = W_2^m = \{f: f, Df, \ldots, D^{m-1}f \text{ are absolutely continuous and } D^m f \in L_2[0,1]\}, \tag{1.4}$$

which is a Sobolev space (Adams, 1975). Let

$$<f,g>_{W_2^m} = <f,g>_{L_2} + <D^m f, D^m g>_{L_2} .$$

If $m \ge 1$ then W_2^m is a reproducing kernel Hilbert space under this inner product.

EXAMPLE 2 Suppose again that $\beta:[0,1] \to \mathbb{R}$ and let $K:[0,1]^2 \to \mathbb{R}$ be smooth. Let the observation vector have components

$$Y_{ni} = \int_0^1 K(t_{ni},s)\ \beta(s)\ ds + \epsilon_{ni}\ , \text{ for } i=1,2,\ldots,n$$

where the t_{ni}'s and ϵ_{ni}'s are as in Example 1. Y_n and σ_n^2 will also be the same as in Example 1. It is not necessary to take Ω as a reproducing kernel space, but only that the integral operator with kernel K, also denoted by K, is a continuous map from Ω onto a reproducing kernel space so that the operator X_n will be bounded (here $X_n\beta = (\int K(t_{n1},s)\beta(s)ds,\ \ldots,\ \int K(t_{nn},s)\beta(s)ds))$. Thus it is reasonable to suppose that

(1.5) $\quad \Omega = W_2^m$ some $m \ge 0$ and $K:\Omega \to W_2^r$ is bounded , some $r \ge 1$.

EXAMPLE 3 Suppose $X_n\beta = (\beta(t_{n1}),\ \ldots,\ \beta(t_{nn}),\ D\beta(t_{n1}),\ \ldots,\ D\beta(t_{nn})) \in \mathbb{R}^{2n}$, i.e. we have discrete noisy observations of the function and its first derivative. Put $Y_n = \mathbb{R}^{2n}$ as sets and

(1.6) $$<\eta_1,\eta_2>_{Y_n} = n^{-1}\sum_{i=1}^{2n} \eta_{1i}\eta_{2i} .$$

Equation (1.2) will hold if $\epsilon_n = (\epsilon_{n1},\ \ldots,\ \epsilon_{n,2n})$ has mean zero uncorrelated components with variance σ^2, and then $\sigma_n^2 = \sigma^2/n$ again. For this setting, it is necessary to take Ω not only to be a reproducing kernel space, but to additionally require that the derivative evaluations $\beta \mapsto D\beta(t)$ are continuous. For this it suffices to assume that

(1.7) $\Omega=W_2^m$ for some $m\geq 2$.

In each of these examples, typical choices of Ω contain functions f such that $f(t_{ni})=y_{ni}$ for all i, i.e. functions that interpolate the data. Whenever such functions exist in the parameter space, the least squares estimator will be one them. Thus, as $n\to\infty$ the least squares estimator will oscillate wildly and not converge to the true function. This motivates us to seek other estimators.

In Section 2 we describe the MOR estimator, which is consistent provided the smoothing parameter is chosen appropriately. The continuous analog is introduced in Section 3 where it is shown how to approximate the discrete problem. This allows us to compute convergence rates for the MOR estimator. A comparison of the convergence rates for Examples 1 and 3 shows that the derivative data in Example 3 significantly reduces the estimation error over Example 1. Confidence regions for Example 1 are discussed in Section 4. Some other possible applications are indicated in Section 5.

2. THE METHOD OF REGULARIZATION. As discussed in the previous section, the least squares estimator is unsatisfactory for the examples considered. We consider a "sieve" estimator (Grenander, 1981) as follows. Let $T:\Omega\to H$ be a bounded linear operator where H is Hilbertian, and for $\mu\geq 0$ put

$$\Omega_\mu = \{ f\in\Omega : \|Tf\|_H^2 \leq \mu \}.$$

The estimator obtained by minimizing the residual sum of squares $\|y_n-X_n\beta\|_{Y_n}^2$ over $\beta\in\Omega_\mu$ will be consistent provided T is appropiately chosen and μ varies appropriately with n. The Langragian for the constrained minimization problem is

$$(2.1) \qquad L_{n\lambda}(\beta) = \lambda\|T\beta\|_H^2 + \|y_n-X_n\beta\|_{Y_n}^2 ,$$

where the Langrange multiplier λ (also known as the smoothing parameter or regularization parameter) is chosen so the minimizer

$\beta_{n\lambda}$ of $L_{n\lambda}$ satisfies $\|T\beta_{n\lambda}\|_H^2 = \mu$. Thus we have a family of estimators which can be parameterized by either μ or λ, and we use λ since for fixed $\lambda \geq 0$ the estimate is a linear function of the data. $\beta_{n\lambda}$ is called the <u>method of regularization</u> (MOR) estimator. Besides the derivation here (which is not based on any optimality principle for the estimator), one can obtain $\beta_{n\lambda}$ as the solution of a linear minimaxity problem (Speckman, 1977, and Li, 1982), or as the solution of a Bayesian estimation problem (Kimeldorf and Wahba, 1970, and Cox, 1986b).

One can show that the minimal norm solution to the problem of minimizing $L_{n\lambda}$ is given by

$$\beta_{n\lambda} = (\lambda W + U_n)^{\dagger} X_n^* Y_n, \tag{2.2}$$

where $X_n^*: Y_n \to \Omega$ is the adjoint of X_n, the dagger † denotes the Moore-Penrose generalized inverse (Groetsch, 1977), and

$$U_n = X_n^* X_n \ , \ W = T^* T.$$

In the cases we consider, $(\lambda W + U_n)$ is invertible for all n sufficiently large so we may use $(\lambda W + U_n)^{-1}$ in place of the Moore-Penrose genalized inverse, and the solution of the minimization problem is unique. We henceforth dispense with the generalized inverse.

It is instructive to work out a concrete representation for U_n in the setup of each example. For Example 1, let $\xi(t)$ denote the representer of evaluation at t, i.e. $\langle \theta, \xi(t) \rangle_\Omega = \theta(t)$, $\forall \theta \in \Omega$ and $\forall t \in [0,1]$. Then one can easily see that

$$X_n^* \eta = n^{-1} \sum_{i=1}^{n} \eta_i \xi(t_{ni}). \tag{2.3}$$

Now define the empirical distribution of the t_{ni}'s by

$$F_n(t) = n^{-1} |\{i : 1 \leq i \leq n \text{ and } t_{ni} \leq t\}|, \tag{2.4}$$

where $|A|$ denotes the cardinality of a set A. Then from (2.3) we have

$$U_n\theta = \int_0^1 \xi(t)\,\theta(t)\,dF_n(t),\ \forall\theta\in\Omega. \tag{2.5}$$

For Example 2, let $\xi(t)$ be the representer of evaluation of $K\theta$ at t, i.e. $\langle\xi(t),\theta\rangle_\Omega = \int K(t,s)\theta(s)ds$ for $\forall\theta\in\Omega$ and $\forall t\in[0,1]$. Then

$$U_n\theta = \int \xi(t) \int K(t,s)\,\theta(s)\,ds\,dF_n(t). \tag{2.6}$$

Example 3 is a little more complicated. Let $\xi(t)$ be the representer of evaluation at t on W_2^m as in Example 1. One can show that the representer of evaluation of the derivative $D\theta$ at t is $\xi'(t)$, the derivative of ξ as a function $[0,1]\to W_2^m$. (Writing $\xi(t)=Q(t,\cdot)$ where Q is the reproducing kernel, we have that $\xi'(t)=\partial Q(t,\cdot)/\partial t$.) It is easy to check that

$$U_n\theta = \int [\xi(t)\theta(t) + \xi'(t)D\theta(t)]\,dF_n(t). \tag{2.7}$$

These formulae for U_n will be used in the sequel. The reason for our choice of the scaling of Y_n inner product in (1.3) and (1.6) was so that U_n could be expressed as an integral operator with measure F_n.

In order to specify the MOR estimator for the examples it is necessary to choose T. Since $\Omega=W_2^m$ in each case it is convenient to use

$$T = D^m : \Omega \to L_2[0,1], \tag{2.8}$$

so that the "penalty" functional is

$$\|T\theta\|_H^2 = \int [D^m\theta(t)]^2\,dt. \tag{2.9}$$

The MOR estimator for Example 1 will be a smoothing spline (Kimeldorf and Wahba, 1970). Such penalties are also widely for Example 2, see e.g. Wahba (1973, 1977) or Nychka, et. al. (1984).

3. THE CONTINUOUS APPROXIMATION. In the previous section we saw that for each of the examples the operator U_n could be expressed in terms of Ω-valued integrals with respect to dF_n, the empirical distribution of the observation points t_{ni}. We assume that there is a distribution function F having Lebesgue density f bounded away from 0 and ∞ (and satisfying other regularity conditions as needed) such that

$$k_n = \sup_{0\le t\le 1} |F_n(t)-F(t)| \to 0 \text{ as } n\to\infty. \tag{3.1}$$

It is then reasonable to conjecture that U_n can be well approximated by the operator U obtained by replacing F_n by F in (2.5) through (2.7). This is investigated in some depth in Cox(1984b, 1986a).

We indicate the nature of the results that can be obtained. The error vector can be decomposed as follows:

$$\beta_{n\lambda} - \beta = B_{n\lambda}\beta + V_{n\lambda} , \tag{3.2}$$

$$B_{n\lambda} = (\lambda W+U_n)^{-1}U_n - I , \tag{3.3}$$

$$V_{n\lambda} = (\lambda W+U_n)^{-1}X_n^* \epsilon_n . \tag{3.4}$$

Note that $B_{n\lambda}\beta$ is the bias vector, i.e. the deterministic component of the estimation error, and $V_{n\lambda}$ is the random component of the estimation error. One message of Cox(1986a) is that one can replace the appearances of U_n by U in (3.3) to obtain a continuous bias operator which approximates $B_{n\lambda}$ in a variety of operator norms (including $W_2^q \to W_2^p$ operator norms with $0\le p<q<m+1/2$ in the setup of Example 1). Also, one can replace U_n by U in the calculation of $E\|V_{n\lambda}\|^2$ for a variety of norms $\|\cdot\|$. The advantage of these results is that the continuous quantities are easier to

analyze than the corresponding discrete quantities since the dependence on n in the continuous quantities is isolated to λ (we must have $\lambda \to 0$ as $n \to \infty$ to obtain consistency). This allows us to obtain results such as the following, which are proved in Cox(1984a, 1986a).

THEOREM 3.1. *For Example 1 suppose* $\Omega = W_2^m$, $m \geq 2$, *and that* $\lambda \to 0$ *and* $k_n \lambda^{-3/4m} \to 0$ *as* $n \to \infty$. *Then if* $0 \leq p < 2m - 3/2$,

$$(3.5) \qquad E\| V_{n\lambda} \|^2_{W_2^p} = O(n^{-1} \lambda^{-(2p+1)/2m}).$$

Suppose $\beta \in W_2^q$ *for some* q *satisfying* $\max\{1,p\} < q < m+1/2$. *Then*

$$(3.6) \qquad \| B_{n\lambda} \beta \|^2_{W_2^p} = O(\lambda^{(q-p)/m}) \| \beta \|^2_{W_2^q} \, ,$$

uniformly in $\beta \in W_2^q$. *If* $\beta \in W_2^p$ *then*

$$(3.7) \qquad \| B_{n\lambda} \beta \|^2_{W_2^p} = o(1).$$

REMARKS. The bound in (3.5) is tight in the sense that the quantity on the r.h.s. is also a lower bound on the order of the l.h.s. The bound in (3.6) is tight in the sense that

$$\sup \{ \| B_{n\lambda} \beta \|^2_{W_2^p} : \| \beta \|_{W_2^q} \leq \alpha \} = O(\lambda^{(q-p)/m}) \alpha^2,$$

uniformly in $\alpha > 0$. One can extend (3.6) to $q > m+1/2$ provided β satisfies certain boundary conditions (see Cox, 1984a, 1986a, and Rice and Rosenblatt, 1981). However, the extension stops at $q=2m$, except when β is a polynomial of degree $<m$ in which case $B_{n\lambda} \beta = 0$ for $\forall n, \lambda$. Note that the best upper bound on the rate of convergence is

$$(3.8) \qquad E\| \beta_{n\lambda} - \beta \|^2_{W_2^p} = O(n^{-2(q-p)/(2q+1)}),$$

obtained by setting $\lambda = n^{-2m/(2q+1)}$. Compare with Stone(1982) and Nussbaum(1985).

THEOREM 3.2. *In the setup of Example 3, suppose* $F(t)=t$ *for* $0\le t\le 1$ *(i.e. uniform asymptotic design). Assume* $\lambda\to 0$ *and* $k_n\lambda^{-3/(4m-4)}\to 0$ *as* $n\to\infty$. *Then if* $1/2<p<2m-5/2$ *we have*

$$(3.9)\qquad E\|V_{n\lambda}\|^2_{W_2^p} = o(n^{-1}\lambda^{-(2p-3)/(2m-2)}),$$

and for $0\le p<1/2$

$$(3.10)\qquad E\|V_{n\lambda}\|^2_{W_2^p} = o(n^{-1}).$$

If $\beta\in W_2^q$ *with* $\max\{p,2\}<q<m+1/2$ *then*

$$(3.11)\qquad \|B_{n\lambda}\beta\|^2_{W_2^p} = o(\lambda^{(q-p)/(m-1)})\|\beta\|^2_{W_2^q},$$

uniformly in $\beta\in W_2^q$.

REMARKS. The remarks regarding the tightness of the bounds hold here as well (except there is no gain in the rate on the bias after $q=2m-1$). The main difference here is that for $p=0$ we can obtain

$$(3.12)\qquad E\|\beta_{n\lambda}-\beta\|^2_{L_2} = o(n^{-1}),$$

provided $\lambda=o(n^{-(m-1)/q})$, which is achievable if $k_n\to 0$ fast enough. The fact that one can obtain the "usual" (parametric) rate of convergence in a nonparametric problem is due to the fact that derivative data is used in the estimation. This should be kept in mind in practice, as in Schwarz(1979).

4. CONFIDENCE REGIONS. In this section we consider the problem of obtaining (simultaneous) confidence regions for a family of linear functionals $\Lambda\subset\Omega$. Our approach is based on ideas of Knafl,

Sacks, and Ylvisaker (1985). For convenience, we shall use $\langle\cdot,\cdot\rangle$ and $\|\cdot\|$ without subscripts to indicate the inner product and norm of Ω. We assume that for some $\mu>0$ the true β lies in $\Omega_\mu = \{f\in\Omega : \|Tf\|_H \le \mu\}$ as defined in Section 2 above. A $1-\alpha$ confidence region for $\langle\Lambda,\beta\rangle_\Omega$ is given by a map $h(\cdot,y_n):\Lambda\to\mathbb{R}$ such that

$$(4.1) \qquad \inf_{\beta\in\Omega_\mu} P_\beta[\ |\langle\eta,\beta-\beta_n\rangle| \le h(\eta,y_n),\ \forall\eta\in\Lambda\ ] \ge 1-\alpha$$

where $\beta_n = \beta_{n\lambda}$ for some deterministic sequence λ_n and P_β denotes probability assuming β is the true parameter. One drawback with this approach is that one must know μ such that $\beta\in\Omega_\mu$. We shall show below that it is possible to estimate μ consistently and so obtain (4.1) but with a "$\lim_{n\to\infty}$" appearing before the "inf". Note that our interpretation of a confidence region is as an accuracy assessment for the estimates $\langle\eta,\beta_n\rangle$ rather than a general set estimate of the parameter β.

Assuming that the error vector ϵ_n is Gaussian (which is asymptotically accurate, as will be seen), it is in principle possible to calculate probable bounds on the random component of the estimation error $\langle\eta,V_{n\lambda}\rangle$ where $V_{n\lambda}$ is given in (3.4). The difficult part of the confidence region problem is obtaining bounds on the bias $\langle\eta,B_{n\lambda}\beta\rangle$. This is the purpose for assuming $\beta\in\Omega_\mu$, for then we may consider the worst case bias. Put

$$(4.2) \qquad b_n(\lambda,\eta) = \sup\{\ \langle\eta,B_{n\lambda}\beta\rangle : \beta\in\Omega_1\ \},$$

and then

$$\sup\{\ \langle\eta,B_{n\lambda}\beta\rangle : \beta\in\Omega_\mu\ \} = \mu b_n(\lambda,\eta),$$

for all $\mu\ge 0$.

PROPOSITION 4.1. *Suppose the following hold: (a) $U_n = X_n^* X$ is compact, and (b) $W = T^*T$ is such that $(W+U_n)^{-1}\in\mathcal{B}(\Omega)$, the space of* *bounded linear operators on* Ω. *Then the following conclusions hold*: (i) *the null space of* W, *denoted* $N(W)$, *has finite dimension*; (ii) $R(B^*_{n\lambda}) \subset R(W)$, *the range of* W; *and* (iii) $b^2_n(\lambda,\eta) = \langle W^{-1}B^*_{n\lambda}\eta, B_{n\lambda}\eta\rangle$.

PROOF. (i) Let $c = \|(U_n+W)^{-1}\|^{-1}_{B(\Omega)}$, where the norm is the usual $B(\Omega)$-operator norm. Then

$$(4.3) \qquad \|U_n x\| \geq c\|x\|, \quad \forall x \in \Omega.$$

This implies $U_n N(W)$ is a closed subspace by an argument similar to one given in Theorem 4.23 of Rudin (1973). Since $U_n|_{N(W)}$ is also compact, $\dim U_n N(W) < \infty$ by Theorem 4.18(b) of Rudin (1973). Now (b) implies $N(U_n) \cap N(W) = \{0\}$ so U_n is injective on $N(W)$ and hence $\dim N(W) < \infty$.

(ii) We first show that $R(W)$ is a closed subspace of Ω. Now $W|_{N(W)^\perp}$ is injective (where $N(W)^\perp$ denotes the orthogonal complement of the subspace $N(W)$), and $R(W)$ is not closed iff W^{-1} is an unbounded operator $\overline{R(W)} \to N(W)^\perp$ iff $\exists \{x_n\} \subset N(W)^\perp$ such that $\|x_n\| = 1$ $\forall n$ but $Wx_n \to 0$. After passing to a subsequence $\{x_n\}$ we have by (a) that Ux_n converges to some z, so $(W+U)x_n \to z$ which implies by (b) that $x_n \to (U+W)^{-1}z$ which implies $z=0$. But $\|(W+U)x_n\| > c\|x_n\| = c > 0$, a contradiction. Hence $R(W)$ is closed.

Next we show $R(W) = N(W)^\perp$. A self-adjointness argument shows $R(W) \subset N(W)^\perp$. For the reverse inclusion, let P be orthogonal projection onto $R(W)$, and $Q = I - P$. Let $v \in N(W)^\perp$, then $\forall u \in \Omega$, $0 = \langle Qv, Wu\rangle = \langle WQv, u\rangle$ which implies $Qv \in N(W)$. Now $P(v) \in R(W) \subset N(W)^\perp$ and $v \in N(W)^\perp$ so $Pv - v = Qv \in N(W) \cap N(W)^\perp = \{0\}$. Hence $Pv = v$ and $v \in R(W)$.

Now it is easy to see that $N(B_{n\lambda}) = N(W)$ and since $R(B^*_{n\lambda}) \subset N(B_{n\lambda})^\perp$ (a general fact), we have $R(B^*_{n\lambda}) \subset R(W)$.

(iii) Since $B_{n\lambda}\varsigma = 0$ for $\varsigma \in N(W)$, in the maximization problem (4.2) neither the objective $\langle B_{n\lambda}\beta, \eta\rangle$ nor the constraint $\langle W\beta, \beta\rangle \leq 1$ is affected by adding to β an element of $N(W)$. Hence we may restrict attention to $\beta \in N(W)^\perp$, on which $\beta \mapsto \langle \beta, W\beta\rangle^{1/2}$ is a norm. Letting $\xi = W^{1/2}\beta$ the maximization problem (4.2) is equivalent to maximizing $\langle B_{n\lambda} W^{-1/2}\xi, \eta\rangle$ subject to $\xi \in W^{-1/2}N(W)^{\perp}$ and $\|\xi\| \le 1$. It is elementary to show the solution is

$$\xi = W^{-1/2} B^*_{n\lambda}\eta \; / \; \| W^{-1/2} B^*_{n\lambda}\eta \| ,$$

which when substituted back into the objective yields the formula for $b_n(\lambda, \eta)$. This completes the proof when $\eta \notin N(W)$. When $\eta \in N(W)$ the result is trivial.

Q.E.D.

In the proof we tacitly assumed $R(W^{-1}) = R(W)^{\perp}$, which is always permissable. Any choice of $W^{-1}B^*_{n\lambda}\eta$ will give the same numerical value for the r.h.s. of the formula in (iii).

The conditions (a) and (b) of the Theorem are satisfied for Example 1. One may also replace U_n by U in the Theorem and the result is still true, and applies to Example 1. In general it is not easy to compute $b_n(\lambda, \eta)$, but it is possible to compute the corresponding continuous analog $b(\lambda, \eta)$ obtained by replacing U_n by U (i.e. $B_{n\lambda}$ by B_λ) in (4.2).

Assuming $N(U) = \{0\}$ we can find eigenvectors $\{\varphi_i : i = 1, 2, \ldots\}$ spanning Ω such that $\langle \varphi_i, U\varphi_j\rangle = \delta_{ij}$ and $\langle \varphi_i, W\varphi_j\rangle = \gamma_i \delta_{ij}$ for all i, j where δ_{ij} is Kronecker's delta and $0 \le \gamma_1 \le \gamma_2 \le \ldots$ are the associated eigenvalues. See Proposition 2.2 of Cox (1986a). Assuming that the asymptotic design is uniform (i.e. $F(t) = t$, $0 \le t \le 1$) then for Example 1 the φ_i's are eigenfunctions of the differential operator $(-D^2)^m$ acting on the subspace of W_2^{2m} of functions satisfying the "natural boundary conditions" (NBC): $h^{(j)}(0) = h^{(j)}(1) = 0$ for $m \le j \le 2m-1$, i.e. $(-D^2)^m \varphi_j = \gamma_j \varphi_j$, the φ_j's satisfy the NBC, and $\{\varphi_j\}$ is a complete orthonormal system for $L_2[0,1]$ (see Cox, 1984a).

Utilizing this eigensystem one may show the the continuous analog of the maximization problem in (4.2) may be expressed as

$$b(\lambda, \eta) = \sup \{ \sum_i \lambda\gamma_i (1+\lambda\gamma_i)^{-1} \langle \beta, U\varphi_i\rangle \langle \eta, \varphi_i\rangle \; : \; \sum_i \gamma_i \langle \beta, U\varphi_i\rangle^2 \le 1 \}$$

for which the solution is

$$b^2(\lambda, \eta) = \sum \lambda^2 \gamma_i (1+\lambda\gamma_i)^2 \langle \eta, \varphi_i\rangle^2 . \tag{4.4}$$

Now we specialize to the case $\eta = \xi(t)$, and write $b(\lambda,t)$ for $b(\lambda,\xi(t))$, and similarly for $b_n(\lambda,t)$. Then

$$b(\lambda,t) = \sum \lambda^2 \gamma_i (1+\lambda\gamma_i)^{-2} \varphi_i^2(t).$$

An alternate form for $b(\lambda,t)$ may be more convenient. Let $G_\lambda(s,t)$ denote the Green's function for the differential operator $\lambda(-D^2)^m+1$ with NBC. One can show

$$G_\lambda(s,t) = \sum (1+\lambda\gamma_i)^{-1} \varphi_i(s)\varphi_i(t),$$

and thus,

$$b^2(\lambda,t) = \lambda[G_\lambda(t,t) - \int G_\lambda^2(t,\tau)d\tau]. \tag{4.5}$$

One possible advantage of (4.5) is that one can obtain closed form expressions for G_λ (albeit with great tedium) and so compute and store once and for all $b(\lambda,t)$ for use in constructing confidence bands. Of course it is necessary to know that $b(\lambda,t)$ is a sufficiently good approximation to $b_n(\lambda,t)$. This is the subject of the following.

Theorem 4.2. *Suppose for Example 1 that the asymptotic design is uniform and that* $\lambda=\lambda_n$ *satisfies (a)* $\lambda\to 0$; *and (b)* $k_n\lambda^{-3/4m} = o(\lambda^\delta)$ *for some* $\delta>0$. *Then*

$$b_n(\lambda,t) = b(\lambda,t) + o([\int b^2(\lambda,\tau)d\tau]^{1/2}),$$

as $n\to\infty$ *uniformly in* $t\in[0,1]$.

Remark. Of course one would prefer to replace the L_2 norm with sup norm in the little-oh above, but we have not found a way of doing this yet, although we conjecture it can be done.

Proof. In the constrained maximization defining $b(\lambda,t)$ we may restrict $\beta\in N(W)^\perp$ on which $\beta \mapsto \langle\beta,W\beta\rangle^{1/2}$ is a norm equivalent to $\|\cdot\|$. Let c be such that $\|\beta\|^2 \le c\langle\beta,W\beta\rangle$, $\forall\beta\in N(W)^\perp$. Then

$$d_n(\lambda) = \sup \{ \langle (B_\lambda - B_{n\lambda})\beta, \xi(t) \rangle : \langle \beta, W\beta \rangle \le 1 \text{ and } 0 \le t \le 1 \}$$
$$\le \sup \{ \langle (B_\lambda - B_{n\lambda})\beta, \xi(t) \rangle : \|\beta\|^2 \le c \text{ and } 0 \le t \le 1 \}.$$

Let $\|\cdot\|_\rho$ denote the ρ-norms defined above Theorem 2.3 of Cox (1986a) or in Definition 2 of Cox (1984b), and let Ω_ρ denote the corresponding space. Then by Lemma 5.1 of Cox (1984b) we have $|\langle (B_\lambda - B_{n\lambda})\beta, \xi(t)\rangle| \le \|B_\lambda - B_{n\lambda}\|_{\mathcal{B}(\Omega, \Omega_{(1+\delta)/2m})} \|\beta\| \; \|\xi(t)\|_{2-(1+\delta)/2m}$ for any $\delta \in (0, 2m-1)$. Here, the norm on $B_\lambda - B_{n\lambda}$ is the usual norm for operators from $\Omega \to \Omega_{(1+\delta)/2m}$. Using Theorems 4.3 and 2.3(c) of Cox (1986a) along with (a) and (b) above one can show that for suitable choice of δ and $\epsilon > 0$,

$$\|B_\lambda - B_{n\lambda}\|_{\mathcal{B}(\Omega, \Omega_{(1+\delta)/2m})} = o(\lambda^\epsilon \|B_\lambda\|_{\mathcal{B}(\Omega, \Omega_{(1+\delta)/2m})})$$
$$= o(\lambda^{1-1/2m}).$$

One must in fact use the proof of Theorem 4.3 of Cox (1986a) to get the factor of λ^ϵ in the first estimate above, noting that all estimates in that proof involve a power of $k_n \lambda^{-3/4m}$ times an appropriate norm of B_λ, and (b) above allows insertion of an extra power of λ in these estimates. As it is easy to show that $\|\xi(t)\|_{2-(1+\delta)/2m}$ is bounded uniformly in t, it follows that

$$d_n^2(\lambda) = o(\lambda^{1-1/2m}).$$

Also,

$$\int b^2(\lambda, t)\,dt = \sum \lambda^2 \gamma_i (1 + \lambda\gamma_i)^{-2} \int \varphi_i^2(t)\,dt$$
$$= \sum \lambda^2 \gamma_i (1 + \lambda\gamma_i)^{-2},$$

and one can show that the latter quantity is bounded above and below by constant multiples of $\lambda^{1-1/2m}$, see e.g. Theorem 2.4 of Cox (1986a) and use $\gamma_i \approx i^{2m}$.

Q.E.D.

Now we turn the problem of consistently estimating $\mu^2 = \langle\beta, W\beta\rangle$, where β is the true parameter in this latter expression. Since $\langle\beta-\beta_n, W(\beta-\beta_n)\rangle \le \|W\|_{*(\Omega)}\|\beta-\beta_n\|^2$, it suffices to obtain a consistent estimate of $\|\beta\|^2$ (note that $|\langle\beta, W\beta\rangle^{1/2} - \langle\beta_n, W\beta_n\rangle^{1/2}| \le \langle\beta-\beta_n, W(\beta-\beta_n)\rangle$). Using these facts and the results from Cox (1984a, 1986a) it is easy to show the following result.

THEOREM 4.3. *Assume uniform asymptotic design for Example 1 and that*

$$k_n\lambda^{-3/4m} \to 0, \tag{4.6}$$

$$n^{-1}\lambda^{-(1+1/2m)} \to 0. \tag{4.7}$$

Then $E\langle\beta-\beta_n, W(\beta-\beta_n)\rangle \to 0$ *as* $n\to\infty$.

REMARKS. The requirement (4.6) is not restrictive but there are some interesting difficulties associated with (4.7). If $\beta\in\Omega_\rho$ for any $\rho>1$ (e.g. if $\beta\in W_2^{m+1}$) then the λ giving the optimal rate of convergence will satisfy (4.7). However if only $\beta\in\Omega=\Omega_1=W_2^m$ but $\beta\notin\Omega_\rho$ for any $\rho>1$, then the optimal λ will probably not satisfy (4.7). If (4.7) does not hold then the conclusion of the theorem will not hold since the variance of $\langle\beta-\beta_n, W(\beta-\beta_n)\rangle$ is bounded above and below by constant multiples of the l.h.s of (4.7). In practice it is probably reasonable to believe that β has ϵ more smoothness than just being in W_2^m so that one may use the optimal λ, which can be presumably estimated from the data by the method of generalized cross validation (Craven and Wahba, 1979; Speckman, 1983).

The final issue to be considered is the Gaussian approximation of the random component of the estimation. We are interested in approximation in C[0,1] (i.e. supremum) norm so as to obtain simultaneous confidence bands for $\beta(t)=\langle\beta,\xi(t)\rangle$, $t\in[0,1]$. Such a uniform approximation by a sequence of Gaussian processes is obtainable from Theorem 0.1 of Cox (1984b) under enough moments for the ϵ_i's. (Note that $W_2^{\delta+1/2}$ norm is stronger than C[0,1] norm for any $\delta>0$.) Thus if one constructs $r_n(t)$ such that

$$(4.8)\qquad P\{\ |\beta_n(t)-E\beta_n(t)|\ \le\ r_n(t),\ \forall t\in[0,1]\ \}\ \ge\ 1-\alpha$$

under the assumption that the ϵ_i's are Gaussian, then (4.8) is also correct asymptotically for non-Gaussian ϵ_i's. Finally, the confidence region will be given by $\beta_n(t)\pm h_n(t)$ where

$$(4.9)\qquad h_n(t) = \mu b_n(t) + r_n(t),$$

where we have suppressed the dependence on λ.

5. OTHER POSSIBLE APPLICATIONS. In this section we briefly consider some other potential applications of the theory of approximation of MOR estimators.

We consider the Backus-Gilbert averaging kernel in the integral equation context of Example 2. It is defined implicitly by

$$(5.1)\qquad \int A_{n\lambda}(t,s)\ \beta(s)\ ds = [(\lambda W+U_n)^{-1}X_n^*X_n\beta](t),$$

i.e. $A_{n\lambda}$ is the kernel for the integral transform that maps a $\beta\in\Omega$ to $\hat{\beta}_{n\lambda}$, the "estimate" when the error vector ϵ_n is 0. In order that the bias be small it is desirable for $A_{n\lambda}$ to be close to a Dirac δ-function. Backus and Gilbert developed various measures of "deltaness" which are useful for assessing the bias properties of the method. See O'Sullivan (1985) for further remarks on this point.

From (2.6) it is easy to see that

$$(5.2)\qquad A_{n\lambda}(\cdot\,,s) = \int [(\lambda W+U_n)^{-1}\xi(t)]\ K(t,s)\ dF_n(t),$$

where $\xi(t)$ is the representer on Ω of the functional $\beta \mapsto \int K(t,s)\beta(s)ds$. The obvious continuous analog of $A_{n\lambda}$ is

$$A_\lambda(\cdot\,,s) = \int [(\lambda W+U_n)^{-1}\xi(t)]\; K(t,s)\; dF(t). \tag{5.3}$$

We conjecture that A_λ is a good enough approximation to $A_{n\lambda}$ when n is large and λ is small that one may look at A_λ to assess deltaness of $A_{n\lambda}$.

Another potential application is the obtainment of equivalent kernels for MOR estimators. This has been carried out in the context of Example 1 by Silverman (1984). Briefly, since the estimate $\beta_{n\lambda}(t)$ is a linear function of the data, there exists for each i, $1\le i\le n$, a function $G_{n\lambda}(\cdot\,,t_i):[0,1]\to\mathbb{R}$ such that

$$\beta_{n\lambda}(t) = n^{-1}\sum Y_i G_{n\lambda}(t,t_i).$$

One would naturally that there is a continuous analog $G_\lambda(t,\tau)$ such that

$$\beta_{n\lambda}(t) \cong n^{-1}\sum Y_i G_\lambda(t,t_i).$$

This is indeed correct (depending on one's meaning of "$\cong$"), and in fact for Example 1 the G_λ is given in Section 4 above.

We have assumed in all cases that the domain of the function β is [0,1], which is certainly not always the case. Of course a domain of [a,b], $-\infty<a<b<\infty$, can be reduced to this case. Analogs of many of the results quoted above are obtainable when the domain is a bounded region in $\mathbb{R}^d$ with a sufficiently nice boundary (the Gaussian approximation in Section 4 being an exception). Many of the speakers at the conference have been interested in the case when the domain is the sphere in $\mathbb{R}^3$. Assuming a uniform asymptotic design measure, it should be possible to obtain analogous results for this setup. The eigenfunctions φ_i introduced in Section 4 will be spherical harmonics. The only difficulty seems to be in obtaining estimates like those in Assumption 4.1(f) of Cox (1986a). Once this is done the corresponding results should be similar to those of any domain in $\mathbb{R}^2$.

REFERENCES.

Adams, R. (1975) Sobolev Spaces, Academic Press, New York.

Backus, G. and Gilbert, F. (1968) The resolving power of gross earth data, Geophysics Journal of the Royal Astronomical Society, 266, 169-205.

Backus, G. and Gilbert, F. (1970) Uniqueness in the inversion of gross earth data, Philos. Trans. Royal Soc. Series A, 266, 123-192.

Cox, D. D. (1984a) Multivariate smoothing spline functions, SIAM J. Numer. Anal., 21, 377-403.

Cox, D. D. (1984b) Gaussian approximation of smoothing splines, Technical Report No. 743, Dept. of Statistics, University of Wisconsin, Madison, Wisconsin, submitted to Z. Wahrscheinlichkeitstheorie.

Cox, D. D. (1986a) Approximation of method of regularization estimators, tentatively accepted Ann. Statist.

Cox, D. D. (1986b) An analysis of Bayesian inference for nonparametric regression, in preparation.

Cox, D. D. and Nychka, D. (1984) Convergence rates for regularized solutions of integral equations form noisy data, tentatively accepted Ann. Statist.

Craven, P. and Wahba, G. (1979) Smoothing noisy data with spline functions: estimating the correct degree of smoothing by the method of generalized cross validation, Numer. Math., 31, 377-403.

Grenander, U. (1981) Abstract Inference, John Wiley and Sons, New York.

Groetsch, C. (1977) Generalized Inverses of Linear Operators, Marcel-Dekker, New York.

Kimeldorf, G. and Wahba, G. (1970) A correspondence between Bayesian estimation on stochastic processes and smoothing by splines, Ann. Math. Statist., 41, 495-502.

Knafl, G., Sacks, J., and Ylvisaker, D. (1985) Confidence bands for regression functions, J. Amer. Statist. Assoc., 80, 683-691.

Li, K. C. (1982) Minimaxity of the method of regularization on stochastic processes, Ann. Statist., 10, 937-942.

Nadaraya, E. (1964) On estimating regression, Theory Prob. Appl., 9, 141-142.

Nussbaum, M. (1985) Spline smoothing in regression models and asymptotic efficiency in L_2, Ann. Statist., 13, 984-997.

Nychka, D., Wahba, G., Goldfarb, S., and Pugh, T. (1984) Cross-validated spline methods for the estimation of three-dimensional tumor distributions from observations on two-dimensional cross sections, J. Amer. Statist. Assoc., 79, 832-846.

Rice, J. and Rosenblatt, M. (1981) Integrated mean square error of a smoothing spline, J. Approx. Theory, 33, 353-369.

Rudin, W. (1973) Functional Analysis, McGraw-Hill, New York.

O'Sullivan, F. (1985) Inverse problems: bias, variability, and the selection of smoothing parameters, manuscript, Dept. of Statistics, Univ. of Calif. Berkeley.

Schwarz, K. (1979) Geodetic improperly posed problems and their regularization, Bolletino di Geodesia e Scienze Affini, 38, 389-416.

Silverman, B. (1984) Spline smoothing: the equivalent variable kernel method, Ann. Statist., 12, 898-916.

Speckman, P. (1979) Minimax estimates of linear functionals on a Hilbert space, unpublished manuscript.

Speckman, P. (1983) Efficient nonparametric regression with cross validated smoothing splines, submitted to Ann. Statist.

Stone, C. (1982) Optimal global rates of convergence for nonparametric regression, Ann. Statist., 10, 1040-1053.

Tikhonov, A. (1963) Solution of incorrectly posed problems and the method of regularization, Soviet Math. Dokl., 5, 1035-1038.

Wahba, G. (1973) Convergence rates of certain approximate solutions to Fredholm integral equations of the first kind, J. Approx. Theory, 7, 167-185.

Wahba, G. (1977) Practical approximate solutions to linear operator equations when the data are noisy, SIAM J. Numer. Anal., 14, 651-667.

Watson, G. (1964) Smooth regression analysis, Sankhya, series A, 26, 359-372.

DEPARTMENT OF STATISTICS
UNIVERSITY OF ILLINOIS
URBANA, ILLINOIS 61801

Contemporary Mathematics
Volume 59, 1986

PARTIAL SPLINE MODELLING OF THE TROPOPAUSE AND OTHER DISCONTINUITIES

Grace Wahba [1]

ABSTRACT. We show how surfaces (distributions) in two and three dimensions which are smooth except for specified types of discontinuities may be modelled with the use of partial splines. Using the partial spline model one may then estimate the distribution given scattered, noisy, direct or indirect observations, and the resulting estimate will possess the assumed type of discontinuity. The motivation for this work is the estimation of the three dimensional atmospheric temperature distribution given direct and indirect measurements of temperature and the location of the tropopause. The tropopause is modelled as jump of known location (but unknown size) in the vertical first derivative. Side information in the form of linear and nonlinear inequality constraints may be incorporated in the estimate. The approach described here should be applicable to other two and three dimensional imaging problems when the unknown distribution has discontinuities of known location but unknown magnitude.

1. INTRODUCTION. Consider an observer rising through the atmosphere on a weather balloon. In general the temperature will more or less smoothly decrease, until, somewhere in the general vicinity of 25,000-45,000 feet, depending on latitude, longitude and time of year, the observer will experience an abrupt stop to the decline, and the temperature will begin to increase. The location of this sharp minimum is known as the tropopause. Similarly, an observer sinking in the ocean will observe the temperature decrease more or less smoothly, and then, at the thermocline, abruptly stop decreasing and begin to rise. The problem we are considering is the modelling of the three dimensional atmospheric temperature structure, given the location of the tropopause and direct and indirect measurements of the temperature. It is desired to be able to estimate the three dimensional atmospheric temperature distribution given the kind of data collected by the worldwide weather observing system, for the purpose of preparing input initial conditions for a system of partial differential equations of motion of the atmosphere, whose integration forward results in a weather forecast. For the three dimensional temperature distribution of the ocean, wanted mainly for descriptive purposes, only direct measurements are available, but the mathematical methods we will describe for the atmosphere appear to be applicable to the ocean also.

Direct measurements of the upper atmospheric temperature consist of temperature readings from the global radiosonde (balloon) network, and indirect measurements from satellites of upwelling radiation at various frequencies, which are related to the vertical temperature profile by the (mildly nonlinear) equations of radiative transfer.

1980 Mathematics Subject Classification 45L10, 65D07, 65D10.

[1] This work supported by ONR Contract N00014-77-C00675 and NASA Grant 5-316.

In this paper we will review results from Wahba (1985b), O'Sullivan and Wahba (1985), Shiau, Wahba, and Johnson (Dec. 1985) , and Wahba (1984a), and note how by putting them all together we have an approach to solving multidimensional remote sensing problems when the unknown function it is desired to recover (the three dimensional temperature distribution) is smooth except for a particular type of discontinuity whose location is known. It is believed that this approach has applications to other two and three dimensional imaging problems where it is desired to recover an otherwise smooth distribution from direct and/or indirect measurements when the location and nature of the discontinuity (but not necessarily its magnitude) are known.

2. PARTIAL SPLINE MODELS. Partial spline models are models of some response as the sum of a function of one or several variables which is only known to be "smooth" in some sense, plus a parametric function of specified functional form with a small number of unknown parameters. For a description of a variety of multivariate partial spline functions and further references, see Wahba (1985c), Wahba (1984a), Wahba (1984b). (There is a growing recent literature on univariate partial spline models, see the references in Wahba (1985c) . The partial spline models we will be concerned with here model a "response" (here temperature) as a smooth function of latitude, longitude and pressure (the vertical coordinate) plus a function with a discontinuity in the vertical first derivative at a known value of pressure, which may vary with latitude and longitude. For a more general theory of partial spline models involving various types of discontinuities, see Shiau (June, 1985). We will only concern ourselves here with a particular type of discontinuity which appears to describe the tropopause and the thermocline in a reasonable way.

Let Ω be a region of interest in the atmosphere, t a point in Ω, and we will let $t=(P,z)$ where P is a pair of horizontal coordinates and z is the vertical coordinate. Define the "tropopause break function" $\gamma(P,z)$ by

$$\gamma(P,z)=|z-z^*(P)| \tag{2.1}$$

where z* is a known (smooth) function of P. By "smooth" here is meant is a member of a reproducing kernel Hilbert space with a specified number of continuous derivatives. The atmosphere f is modelled as

$$f(t)=g(t)+\theta(P)\gamma(P,z) \tag{2.2}$$

where we will assume that g is in a reproducing kernel Hilbert space H with at least continuous first derivatives. θ may be a constant, or a parametric function of the form:

$$\theta(P)=\sum_{q=1}^{q=r}\theta_q\psi_q(P). \tag{2.3}$$

Then

$$\left.\frac{\partial f}{\partial z}\right]_{z=z^*_-}-\left.\frac{\partial f}{\partial z}\right]_{z=z^*_+}=2\theta.$$

Our model for the data is

$$Y_i=L_i(f)+\varepsilon_i,\ i=1,\dots,n \tag{2.4}$$

where the ε_i's are independent, identically distributed zero mean random variables with common, possibly unknown variance, and L_i is a bounded linear functional defined on the Hilbert space which is the direct sum of H and H_{00}, H_{00} being the r dimensional space spanned by the functions γ_q defined by

$$\gamma_q = \psi_q \gamma. \tag{2.5}$$

The L_i's may be evaluation functionals ($L_i = f(t_i)$), or may, for example be line integrals ($L_i f = \int K_i(t) f(t) dt$). f is estimated by finding g in H and $\theta = (\theta_1, \ldots, \theta_r)'$ to minimize

$$\frac{1}{n} \sum_{i=1}^{i=n} (Y_i - L_i(g + \sum_{q=1}^{q=r} \theta_q \gamma_q))^2 + \lambda J(g). \tag{2.6}$$

where the penalty functional J is a seminorm on H with $m<n$ dimensional null space. The choice of J will be discussed later.

Let H_0 be the null space of J in h and let $\phi_1, \ldots, \phi_M$ span H_0. Let $Q(s,t), s, t \varepsilon \Omega$ be the reproducing kernel for H. Let $\xi_j(t) = L_{j(s)} Q(t,s)$ where $L_{j(s)}$ means that L_j is to be applied to what follows considered as a function of s. Then it can be shown that if there is a unique minimizer of (2.6) then g must be of the form

$$g = \sum_{j=1}^{j=n} c_j \xi_j + \sum_{\nu=1}^{\nu=M} d_\nu \phi_\nu. \tag{2.7}$$

A proof may be constructed following Kimeldorf and Wahba (1971), for further details see Wahba (1984a).

Let T be the $n \times M$ matrix with $i\nu$th entry $L_i \phi_\nu$, let T_1 be the $n \times r$ matrix with iqth entry $L_i \gamma_q$ and let Σ be the $n \times n$ matrix with ijth entry $L_{i(s)} L_{j(s)} Q(s,t) = L_i \xi_j = \langle \xi_i, \xi_j \rangle$ where $\langle .,. \rangle$ is the inner product in $H - H_0$. If $(T:T_1)$ has rank $M+r$ and Σ has rank n, then (2.6) will have a unique minimizer and the minimization of (2.6) reduces to finding c, d, and θ to minimize

$$\frac{1}{n} \| Y - \Sigma c - Td - T_1 \theta \|^2 + \lambda c' \Sigma c \tag{2.8}$$

where $Y = (Y_1, \ldots, Y_N)'$, $c = (c_1, \ldots, c_n)'$, and $d = (d_1, \ldots, d_M)'$. The estimate f_λ of f,

$$f_\lambda = \sum_{j=1}^{j=n} c_j \xi_j + \sum_{\nu=1}^{\nu=M} d_\nu \phi_\nu + \sum_{q=1}^{q=r} \theta_q \gamma_q \tag{2.9}$$

may be shown to be a Bayes estimate with a certain improper prior, following the same arguments as in Wahba (1978).

For very large data sets, it will be desireable to minimize (2.6) for g not in H but in some suibably chosen subspace of H or H_1. For concreteness we will suppose that this subspace is of the form span $B_1, \ldots, B_N, \phi_1, \ldots, \phi_M$, with the B_j's in H_1. Then

$$f_\lambda = \sum_{j=1}^{j=N} c_j B_j + \sum_{\nu=1}^{\nu=M} d_\nu \phi_\nu + \sum_{q=1}^{q=r} \theta_q \gamma_q \tag{2.10}$$

and we must find c, d and θ to minimize

$$\frac{1}{n} \| Y - Xc - Td - T_1\theta \|^2 + \lambda c'Jc \tag{2.11}$$

where T and T_1 are as before, X is the $n \times N$ matrix with ij th entry $L_i B_j$ and J is the $N \times N$ matrix assumed to be of full rank with ij th entry $<B_i, B_j>$. It is an interesting approximation theoretic question how to choose the B_j. Assuming that the goal is to obtain a good but cheap approximation to the minimizer of (2.6), it is desireable that the B_j have good ability to approximate any element in some ellipsoid in H which is belived to contain the solution, (more precisely, any ellipsiod in the span of the ξ_j's). In addition it is desireable that the B_j have compact support for then the large matrices in (2.11) will be sparse and it will be easier to solve for c, d, and θ, and (as we shall discuss later) the generalized cross validation (GCV) estimate of λ. The first N eigenfunctions of Q are known to have good approximation theoretic properties (see Wahba and Micchelli (1981)), and then J will be diagonal. However in this case X will generally be full, and unless H_1 is a space of periodic functions the eigenfunctions are generally not known. There is also some evidence that a set $Q_{s_1}, \ldots, Q_{s_N}$ of representers of evaluation in H_1, (i. e. $<Q_{s_i}, f> = f(s_i)$), for suitably chosen s_i have good approximation theoretic properties in some cases. (see, e. g. Melkman and Micchelli (1978)).

If Ω is a subset of the real line and $H = W_2^m$, certain linear combinations of the Q_{s_i} have compact support (one can get the B-splines this way), but in higher dimensions it is part of the folklore that basis functions with compact support cannot be constructed from span $\left\{ Q_{s_i}, \ldots, Q_{s_N} \right\}$ for H one of the usual Sobolev spaces.

3. THIN PLATE AND OTHER PENALTY FUNCTIONALS. A popular "general purpose" penalty functional in d dimensions is the so called "thin plate" penalty functional. Letting $t = (x_1, \ldots s_d)$, $g(t) = g(x_1, \ldots, x_d)$, we have

$$J(g) = \sum_{\alpha_1 + \ldots + \alpha_d} \frac{m!}{\alpha_1! \cdots \alpha_d!} \int \cdots \int \left[\frac{\partial^m g}{\partial x_1^{\alpha_1} \cdots \partial x_d^{\alpha_d}} \right]^2 dx_1 \cdots dx_d. \tag{3.1}$$

This is an isotropic penalty functional but can be rendered elliptical by measuring $x_1, \ldots, x_d$ in different units. A so called semi-kernel for H_1 is known and has a particularly simple form (see Duchon (1976), Wahba and Wendelberger (1980) and references cited there). This penalty functional is the simplest penalty functional in a (large) equivalence class of penalty functionals. It can be argued that from a numerical analytic /statistical point of view which member of an equivalence class is used does not affect the theoretical solution (much), hence one might as well use that member of the class that is easiest to compute. See the remarks in Section 2 of Wahba (1981) and also Stein (Nov. 85). In the ordinary thin plate spline case ($\theta = 0$) Hutchinson (April 1984) has developed transportable code based on the thin plate spline basis functions in Wahba (1980b) which are essentially linear combinations of the Q_{s_i}. See also Bates et al. (November 1985). It would be very interesting to develop basis functions with compact support which provided a good approximation to the span of a set of Q_{s_i}'s in the thin plate case. For a discussion of the choice of J based on historical data, see Wahba (1982a).

4. DATA BASED CHOICE OF λ AND OTHER PARAMETERS. The GCV estimate of λ is the minimizer of $V(\lambda)$ defined by

$$V(\lambda) = \frac{\| (I - A(\lambda))Y \|^2}{\left[\frac{1}{n} Tr (I - A(\lambda) \right]^2} \tag{4.1}$$

where $A(\lambda)$ is the influence matrix, which is defined by

$$\begin{bmatrix} L_1 f_\lambda \\ \cdots \\ L_n f_\lambda \end{bmatrix} = A(\lambda) Y. \tag{4.2}$$

This estimate for λ has been shown to have good properties in a variety of circumstances, see Wahba (1985a) and references cited there. Certain other parameters, such as m in (3.1) (Wahba and Wendelberger, 1980) or a relative scale factor on one of the x_j's (Wendelberger, 1982) can also be chosen by GCV. One can fix, say m, find the GCV estimate of λ for that m, and then compare the V's for the different m's. It is not suggested, however that parameters which do not correspond to distinct equivalence classes of penalty functionals be chosen this way.

5. NONLINEAR FUNCTIONALS. In the recovery of atmospheric temperature from indirect measurements the indirect measurements are satellite observed radiances which are actually available in the form of mildly nonlinear functionals. The model is :

$$Y_i = N_i f + \varepsilon_i, \tag{5.1}$$

where

$$N_i f = \int K_i(t, f(t)) dt. \tag{5.2}$$

where the Y_i's are satellite observed radiances and the K_i's are derived from the equations of radiative transfer and the transfer properties of the individual channels on the satellite borne radiometer. Then L_i in Eq. (2.6) is replaced by N_i. The resulting nonquadratic optimization problem with a trial value of λ is solved by solving a sequence of quadratic optimization problems. The GCV estimate of λ is based on Eq. (4.1) computed for the quadratic approximation to the problem at convergence. See O'Sullivan and Wahba (1985), Wahba (1985b). Svensson has recently applied a variant of the method in O'Sullivan and Wahba with a break at the tropopause, in one dimension.

6. NON GAUSSIAN ERRORS. If the ε_i's in Eq. (2.4) are normally distributed, it is natural that the first term in (2.6) be a sum of squares. If the ε_i's have some other distribution(s), this term may be replaced by a log likelihood. See O'Sullivan (1983), O'Sullivan, Yandell, and Raynor (1984).

7. LINEAR INEQUALITY CONSTRAINTS. Equation (2.6) may be minimized subject to side conditions which can be represented in terms of linear equality constraints (Wahba, 1980a) or inequality constraints (Villalobos and Wahba, March 1985), (Wahba, 1982b).

8. AN EXAMPLE OF A TROPOPAUSE BREAK FUNCTION. In this Section we present figures to demonstrate the appearance of a hypothetical, but realistic "tropopause break function" of Eq. (2.1) restricted to one horizontal (P) and one vertical (z) coordinate. Fig. 8.1 shows a hypothetical tropopause $z^*(P)$ and Fig. 8.2 gives the associated break function $\gamma(P,z) = |z - z^*(P)|$. In accordance with meteorological practice the figure has been tipped on its side (from a mathematician's point of view). Fig. 8.3 gives a simulated two dimensional model temperature field, constructed with the aid of the tropopause of Fig. 8.1, and intended to simulate a vertical slice of a realistic atmospheric temperature distribution. Fig. 8.4 gives 10 equally spaced sections of the surface of Fig. 8.3, schematically displaced by equal amounts (marked "true" in the figure). The circles represent simulated noisy (direct) observations from the "true" distribution. Fig. 8.5 represents the estimated temperature distribution, and the dashed lines in Fig. 8.4 are the corresponding sections from Fig. 8.5. The surface in Fig. 8.5 was obtained by minimizing Eq. (2.6) with the $n = 150$ data points represented by the circles from Fig. 8.4 with $q = 2$, $\psi_1 = 1$, $\psi_2 = P$, with $J(g)$ given by Eq. (3.1) with $d = 2$, $m = 2$, and using the GCV estimate of λ.

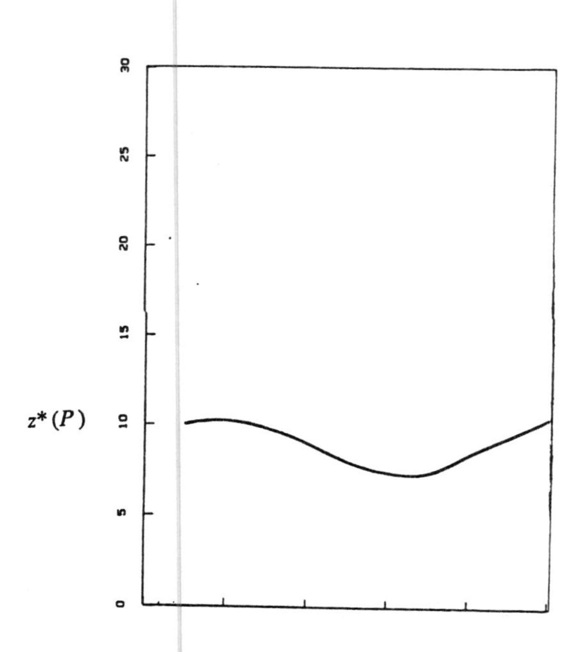

Fig. 8.1. The tropopause, $z^*(P)$.

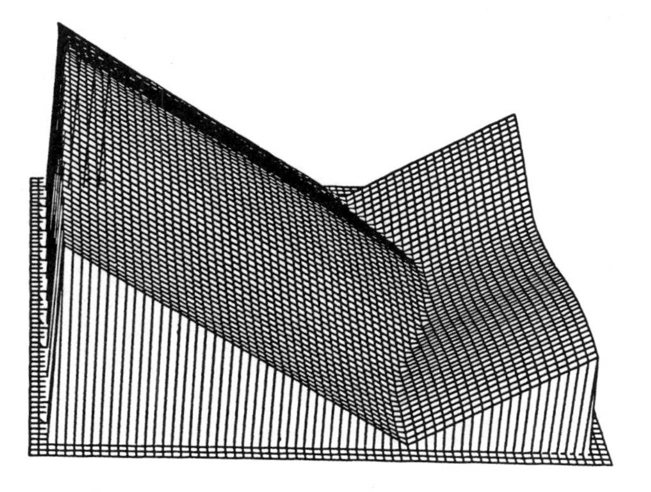

Fig. 8.2. The tropopause break function $\gamma(z, P)$.

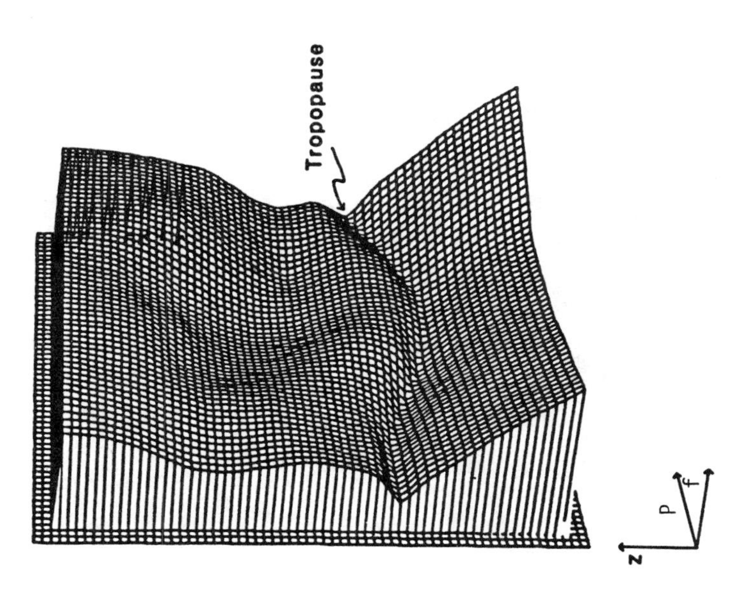

Fig. 8.3. True (simulated) temperature, as a function of z and P.

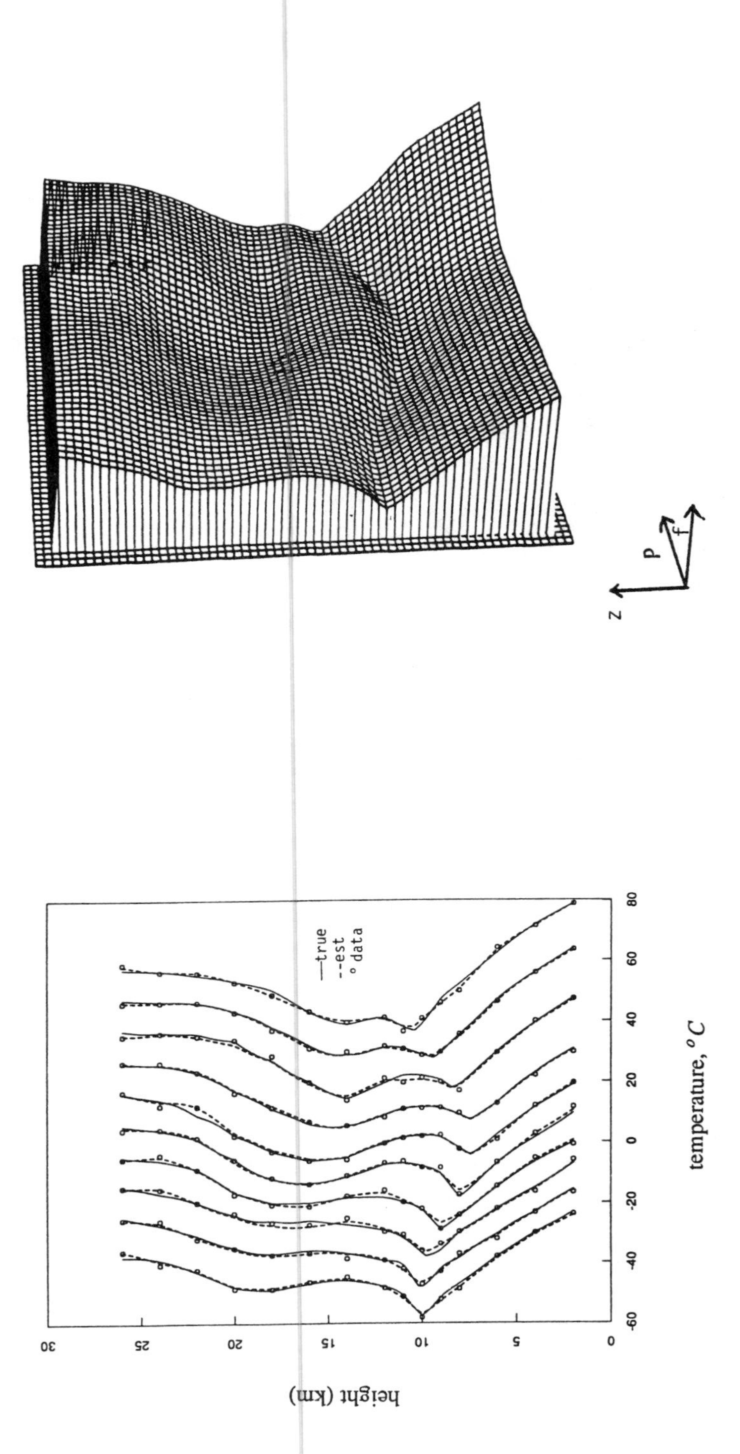

Fig. 8.4. True (simulated) temperatures, simulated data, and estimated temperature curves, with a break in the first derivative, for 10 hypothetical observing stations equally spaced along a fixed latitude.

Fig. 8.5 Estimated temperature distribution given the tropopause height $z^*(P)$ and the data (circles) from Fig. 8.4.

REFERENCES

Bates, D.M., Lindstrom, M.J., Wahba, G., and Yandell, B. (November 1985) ''GCVPACK - Routines for Generalized Cross Validation.'' Technical Report 775, Dept. of Statistics, University of Wisconsin-Madison.

Duchon, J. (1976), ''Fonctions-Spline Et Esperances Conditionnelles De Champs Gaussiens,'' *Annales Scientifiques de L'Universite de Clermont*, 14, 19-27.

Hutchinson, M.F. (April 1984), *A Summary of Some Surface Fitting and Contouring Programs for Noisy Data*, Canberra, Autralia: CSIRO Division of Mathematics and Statistics. (Consulting Report No. ACT 84/6)

Kimeldorf, G., and Wahba, G. (1971), ''Some Results on Tchebycheffian Spline Functions,'' *J. Math. Anal. Applic.*, 33, 82-95.

Melkman, A.A., and Micchelli, C.A. (1978), ''Spline Spaces are Optimal for L 2 N-Widths,'' *Illinois Journal of Mathematics*, 22, 541-564.

O'Sullivan, F. (1983), *The Analysis of Some Penalized Likelihood Estimation Schemes*, University of Wisconsin-Madison, Statistics Dept.. (Technical Report #726)

O'Sullivan, F., Yandell, B., and Raynor, W. (1984), *Automatic Smoothing of Regression Functions in Generalized Linear Models*, University of Wisconsin-Madison, Dept. of Statistics. (Technical Report #734)

O'Sullivan, F., and Wahba, G. (1985), ''A Cross Validated Bayesian Retrieval Algorithm for Non-Linear Remote Sensing Experiments,'' *J. Comput. Physcs*, 59, 441-455.

Shiau, J., Wahba, G., and Johnson, D.R. (Dec. 1985) ''Partial Spline Models for the Inclusion of Tropopause and Frontal Boundary Information in Otherwise Smooth Two and Three Dimensional Objective Analysis.'' Technical Report # 777 University of Wisconsin-Madison Statistics Dept..

Shiau, J. (June, 1985) ''Smoothing Spline Estimation of Functions with Discontinuities.'' Technical Report # 768 University of Wisconsin-Madison Statistics Dept, Madison, WI:.

Stein, M.L. (Nov. 85) ''Asymptotically Efficient Spatial Interpolation with a Misspecified Covariance Function.'' Technical Report 186, Dept. of Statistics, The University of Chicago, Chicago, ILL:.

Svensson, J. (February, 1985), *A Nonlinear Inversion Method for Derivation of Temperature Profiles from TOVS Data*, Norrkoping, Sweden: The Swedish Meteorological and Hydrological Institute. (manuscript, presented at The Second International TOVS Study Conference)

Villalobos, M., and Wahba, G. (March 1985), *Inequality Constrained Multivariate Smoothing Splines with Application to the Estimation of Posterior Probabilities*, Dept. of Statistics, University of Wisconsin-Madison. (Technical Report No. 756)

Wahba, G. (1978), ''Improper Priors, Spline Smoothing and the Problem of Guarding Against Model Errors in Regression,'' *J. R. Statist. Soc. B*, 40, 364-372.

Wahba, G., and Wendelberger, J. (1980), ''Some New Mathematical Methods for Variational Objective Analysis Using Splines and Cross-Validation,'' *Monthly Weather Review*, 108, 36-57.

Wahba, G. (1980a) ''Ill Posed Problems: Numerical and Statistical Methods for Mildly, Moderately, and Severely Ill Posed Problems with Noisy Data.'' University of Wisconsin-Madison Statistics Department Technical Report No. 595. (To appear in the Proceedings of the International Conference on Ill Posed Problems, M.Z. Nashed, ed.)

Wahba, G. (1980b), ''Spline Bases, Regularization, and Generalized Cross Validation for Solving Approximation Problems with Large Quantities of Noisy Data,'' in *Approximation Theory III*, ed. W. Cheney Academic Press, 905-912.

Wahba, G., and Micchelli, C. (1981), ''Design Problems for Optimal Surface Interpolation,'' in *Approximation Theory and Applications*, ed. Z. Ziegler Academic Press, 329-348.

Wahba, G. (1981), ''Data-Based Optimal Smoothing of Orthogonal Series Density Estimates,'' *Ann. Statist.*, 9, 146-156.

Wahba, G. (1982a), ''Vector Splines on the Sphere, with Application to the Estimation of Vorticity and Divergence from Discrete, Noisy Data,'' in *Multivariate Approximation Theory, Vol. 2*, ed. W. S. K. Zeller Birkhauser Verlag, 407-429.

Wahba, G. (1982b), ''Constrained Regularization for Ill Posed Linear Operator Equations, with Applications in Meteorology and Medicine,'' in *Statistical Decision Theory and Related Topics III, Vol. 2*, eds. S. S. Gupta, and J. O. Berger Academic Press, 383-418.

Wahba, G. (1984a), ''Cross Validated Spline Methods for the Estimation of Multivariate Functions from Data on Functionals,'' in *Statistics: An Appraisal, Proceedings 50th Anniversary Conference Iowa State Statistical Laboratory*, eds. H. A. David, and H. T. David Iowa State University Press.

Wahba, G. (1984b), ''Partial Spline Models for the Semiparametric Estimation of Functions of Several Variables,'' in *Statistical Analyses of Time Series*, Tokyo: Institute of Statistical Mathematics, 319-329. (Proceedings of the Japan U. S. Joint Seminar)

Wahba, G. (1985a), ''A Comparison of GCV and GML for Choosing the Smoothing Parameter in the Generalized Spline Smoothing Problem,'' *Ann. Statist.*, 13, 1378-1402.

Wahba, G. (1985b), ''Variational Methods for Multidimensional Inverse Problems,'' in *Proceedings of the Workshop on Advances in Remote Sensing Retrieval Methods*, ed. A. Deepak University of Wisconsin-Madison Statistics Department Technical Report 755. (to appear)

Wahba, G. (1985c), ''Comments to Peter J. Huber, Projection Pursuit,'' *Ann. Statist.*, 13, 518-521.

Wendelberger, J. (1982), *Smoothing Noisy Data with Multidimensional Splines and Generalized Cross-Validation*, University of Wisconsin-Madison Statistics Dept. PhD. Thesis.

Contemporary Mathematics
Volume **59**, 1986

CHOICE OF SMOOTHING PARAMETER IN DECONVOLUTION PROBLEMS

John A. Rice[1]

ABSTRACT. A simple deconvolution problem is considered and is used to illustrate issues of smoothing parameter choice. In particular, it is shown that for some purposes, ordinary cross-validation and related techniques can be quite unsatisfactory.

1. INTRODUCTION

This paper is concerned with the choice of a smoothing parameter for deconvolution problems, particularly with pointing out that a reasonable choice of a smoothing parameter for making prediction error small may not be reasonable in terms of the estimation error incurred, and vica versa. These terms will be defined below.

I will illustrate the points I wish to make with some simple computer simulations. However, I hope that the statistical community is as cautious in evaluating these results, or the results of any single investigator's simulations, as it is in the evaluation of other experiments of the more traditional variety. Hopefully, we are all aware that between laboratory variation is often far more important than within laboratory variation. My laboratory in this case is a room with three Sun workstations; the within laboratory variation that you will see below is provided by a pseudo-random number generator. Sources of between laboratory variation include possible bugs in code and the selection of examples.

2. THE DECONVOLUTION PROBLEM AND A REGULARIZED SOLUTION

Because of its analytic and computational tractability, I will consider a quite simple deconvolution problem. More complex and realistic problems are treated in other papers in this volume. However, I believe that the phenomena we will see in this simple situation persist in more complicated situations as well. Let f, g, h be smoothly periodic functions on $[0,1]$ and its periodic extension:

$$f(x) = \int_0^1 g(x-y)h(y)dy .$$

Suppose the h is known, g is unknown, and f is observed with error:

$$y_j = f(x_j) + \epsilon_j, \quad j=0,1,\ldots,n-1$$

1980 Mathematics Subject Classification (1985 Revision). 65R20, 65D10, 62G05.
[1]Supported in part by National Science Foundation grant DMS-8401279.

where the x_j are evenly spaced, $x_j=j/n$ and the ϵ_j are uncorrelated errors with 0 mean and variance σ^2. We may wish to estimate the unknown function g (I will call this the estimation problem) or the unknown function f, perhaps in order to predict future observations y (the prediction problem).

It is convenient for both computational and theoretical purposes to use Fourier analysis. If

$$f_k = \int_0^1 e^{-2\pi ikx} f(x)dx ,$$

and similarly for h_k, g_k, then $f_k=h_k g_k$. We will also make use of the finite Fourier transform:

$$y_{kn} = \sum_{i=0}^{n-1} e^{-2\pi ijk/n} y_j$$

$$f_{kn} = \sum_{i=0}^{n-1} e^{-2\pi ijk/n} f(x_j) ,$$

etc. Now $n^{-1} f_{kn} = f_k + r_{kn}$, where r_{kn} is an aliased term of smaller order than f_k. For simplicity, we will neglect these terms below and proceed as if $f_{kn} = nf_k$, etc.

The estimates of g and f to be considered are obtained by the commonly used method of regularization; the estimate of g is the minimizer of

$$\frac{1}{n}\sum_{i=0}^{n-1} [y_i-(h*g)(\frac{i}{n})]^2 + \frac{\lambda}{(2\pi)^4}\int_0^1 [g''(x)]^2 dx .$$

Here, I have used $h*g$ to denote convolution and have included the factor $(2\pi)^4$ for algebraic convenience. Using the discrete Fourier transform and Parseval's relation, and neglecting aliasing as mentioned above, $\hat{g}$ has Fourier coefficients $\hat{g}_k$ which are the minimizers of

$$\sum_{k=0}^{n-1} [\frac{y_{kn}}{n} - h_k g_k]^2 + \lambda \sum_{k=0}^{n-1} k^4 |g_k|^2 ,$$

and are easily seen to be

$$\hat{g}_k = \frac{y_{kn}}{n} \frac{\bar{h}_k}{|h_k|^2+\lambda k^4} .$$

The resulting estimate of f_k

$$\hat{f}_k = \hat{g}_k h_k = \frac{y_{kn}}{n} \frac{|h_k|^2}{|h_k|^2+\lambda k^4} .$$

Both of these estimates are linear in the data, and their bias properties can be examined via the Backus-Gilbert kernel as outlined by O'Sullivan elsewhere in this volume. In particular,

$$E\hat{g}_k = f_k \frac{\bar{h}_k}{|h_k|^2+\lambda k^4}$$

$$= g_k \frac{|h_k|^2}{|h_k|^2+\lambda k^4} ,$$

so that

$$E\hat{g}(x) = \int_0^1 g(x) w_\lambda (x-y) dy ,$$

where w_λ, the Backus-Gilbert kernel, has Fourier coefficients, $w_k = |h_k|^2/(|h_k|^2+\lambda k^4)$. Similarly,

$$E\hat{f}_k = f_k \frac{|h_k|^2}{|h_k|^2+\lambda k^4} ,$$

so that

$$E\hat{f}(x) = \int_0^1 f(x) w_\lambda (x-y) dy ,$$

has the same effective kernel as does $E\hat{g}$. The bias properties of the two estimates are thus rather similar; as we will see below, the variance properties can be vastly different.

3. MEAN SQUARE ESTIMATION AND PREDICTION ERRORS

I will refer to the total mean square error in estimating g as the Mean Square Estimation Error ($MSEE$). From the definition and orthogonality properties of the finite Fourier transform,

$$E(\frac{y_{kn}}{n}) = f_k ,$$

$$Cov(\frac{y_{kn}}{n}, \frac{y_{jn}}{n}) = \frac{\sigma^2}{n} \delta_{jk} .$$

Thus

$$MSEE = \sum_{k=0}^{n-1} |g_k - E\hat{g}_k|^2 + \sum_{k=0}^{n-1} Var(\hat{g}_k)$$

$$= \sum_{k=0}^{n-1} |g_k|^2 \left(\frac{\lambda k^4}{|h_k|^2+\lambda k^4} \right)^2 + \frac{\sigma^2}{n} \sum_{k=0}^{n-1} \frac{|h_k|^2}{(|h_k|^2+\lambda k^4)^2} .$$

Similarly, the total mean square error of the estimate of f, the Mean Square Predicition Error ($MSPE$) is

$$MSPE = \sum_{k=0}^{n-1} |f_k|^2 \left(\frac{\lambda k^4}{|h_k|^2+\lambda k^4} \right)^2 + \frac{\sigma^2}{n} \sum_{k=0}^{n-1} \frac{|h_k|^4}{(|h_k|^2+\lambda k^4)^2} .$$

The crucial difference between the two mean square errors lies in the variance terms - in one case the numerator of the sum is $|h_k|^2$ and in the other case $|h_k|^4$. Since h is a smooth function, $h_k \rightarrow 0$ fairly rapidly, and in the case of estimation error, the parameter $\lambda > 0$ is needed to keep the variance sum from diverging. If $n \rightarrow \infty$, and $\lambda \rightarrow 0$ but not too fast, the estimates are consistent, but converge at different rates. The optimal rate at which $\lambda \rightarrow 0$ is different for estimation error than for prediction error. This can be seen estimating the orders of magnitude of the sums in the expressions for mean square errors under the assumption that the Fourier coefficients decay at algebraic rates, for example.

4. NUMERICAL EXAMPLES

I will present four numerical examples. In all of them $n=256$ and the pseudo-random errors are normal with mean 0 and standard deviation .1 generated by the SLATEC package of numerical algorithms. The examples are qualitatively similar to problems that arise in spectroscopy.

EXAMPLE I. In Example I, g is a normal density with standard deviation 10 and h is a normal density with standard deviation 25. Convolution with h substantially smears out the shape of g, and since the Fourier coefficients of h decrease rapidly, undoing the convolution is difficult.

Figure IA shows the mean square errors as functions of λ, from which it is apparent that the optimal values of λ for estimation and prediction are quite different.

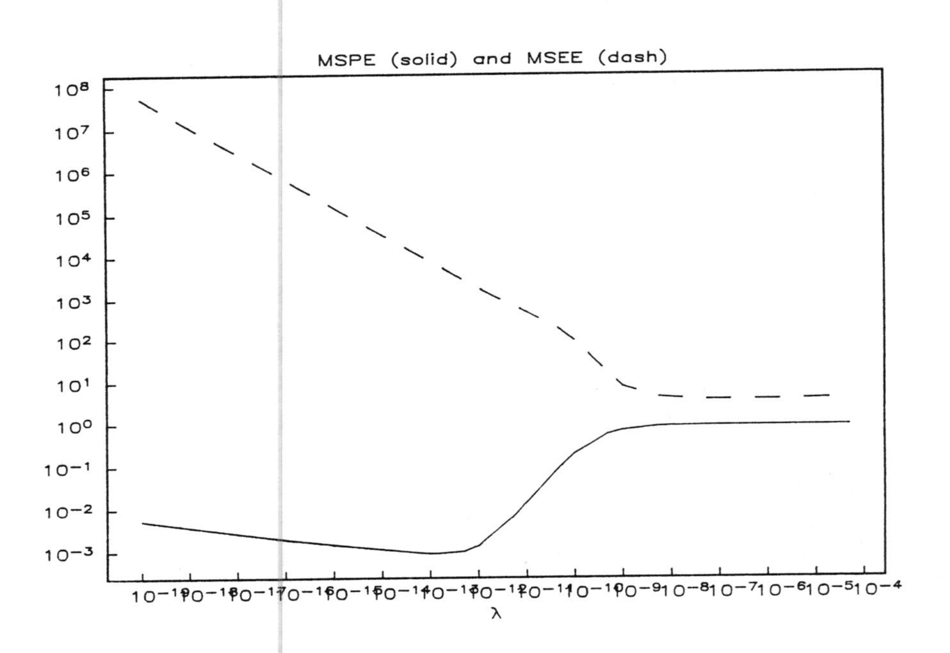

Figure IA.

Figure IB shows, for a single realization, the estimates of f and g constructed with λ chosen optimally for prediction error. The estimate of f looks fine, but g is totally swamped in the noise.

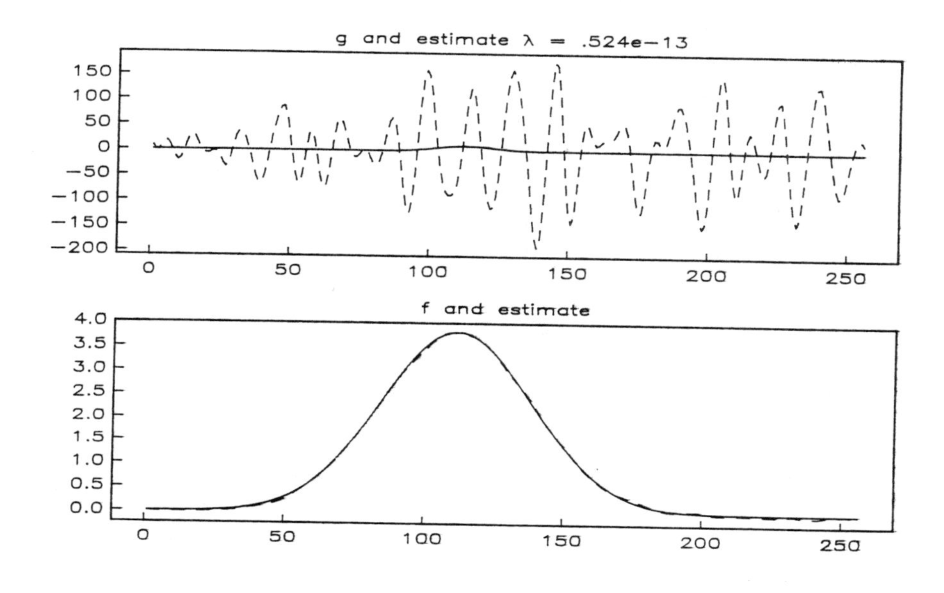

Figure IB.

Figure IC shows estimates of f and g with λ chosen to be optimal for estimating g. The shape of g now begins to emerge from the noise, although the resolution leaves a lot to be desired. Recalling that the standard deviation of the errors is .1, we see that the estimate of f is grossly oversmoothed. David Donoho has developed some theoretical results which show that f must be oversmoothed in order to obtain a consistent estimate of g, in the sense that if the distance between the data and $\hat{f}$ is measured by a chi-square statistic, say, the p-value must tend to 0. In fact, the smoothing parameter is often chosen in practice by using a chi-square measure and a conventional p-value such as .01 or .05. This example suggests that in a hard deconvolution problem, this practice may lead to quite undersmoothed estimates of g.

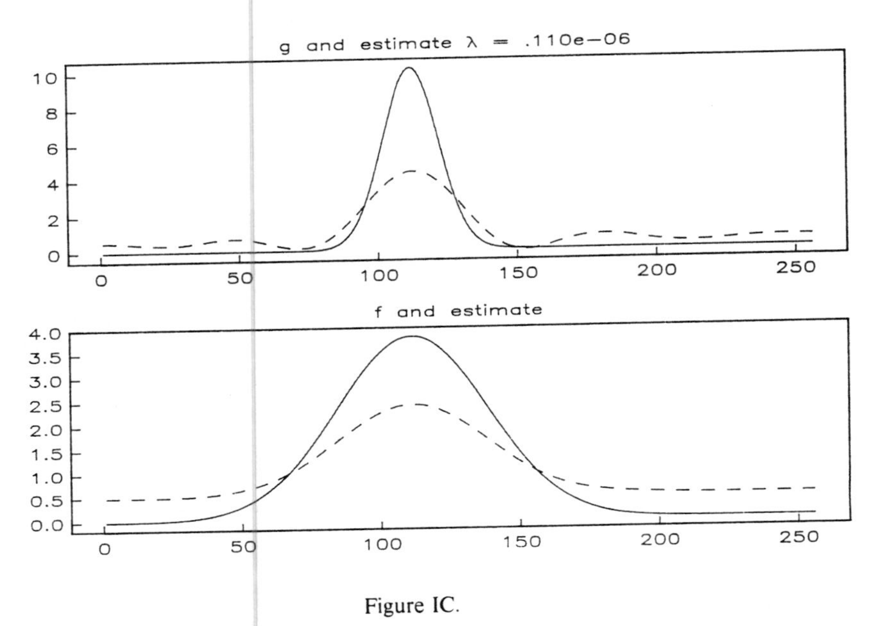

Figure IC.

EXAMPLE II. Figures IIA and IIB show the corresponding results for a realization for which the standard deviation of h is decreased to 5 and the standard deviation of g is still 10 $-\hat{g}$ is thus less smeared out by the convolution than in Example I. Again we see that the same things happen qualitatively. If λ is chosen optimally for prediction, $\hat{g}$ is extremely erratic, and if λ is chosen optimally for estimation of g, $\hat{f}$ is oversmoothed relative to the data variability. (There is an interesting doublet in the estimate of g.)

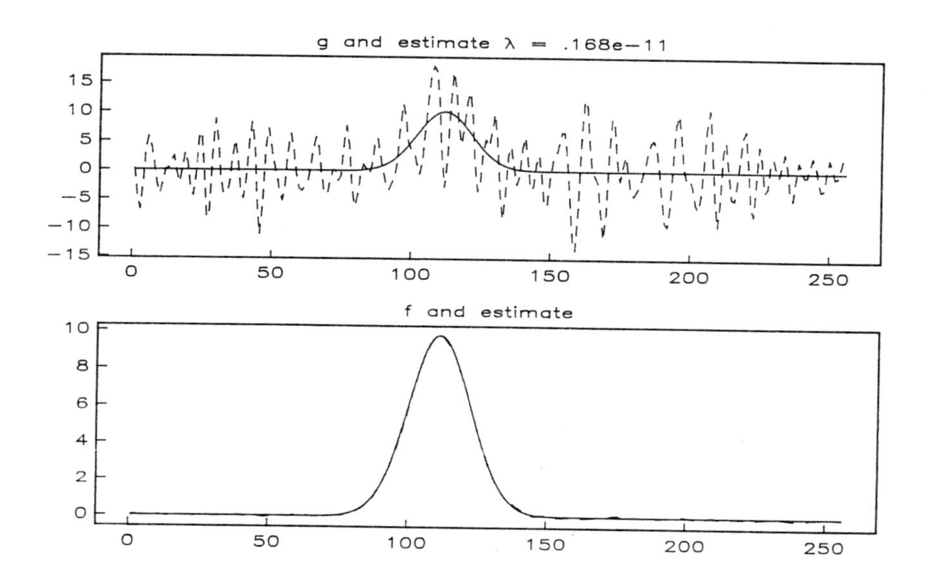

Figure IIA.

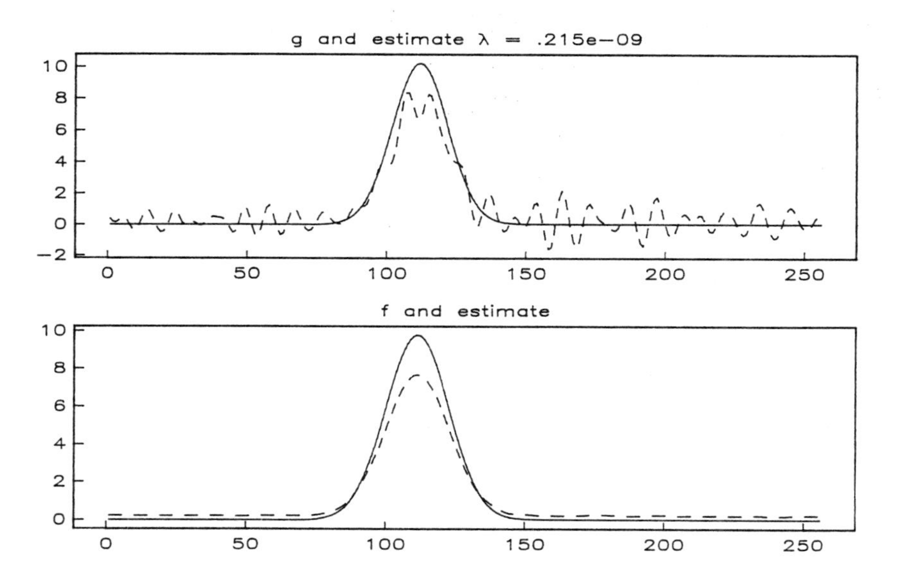

Figure IIB.

EXAMPLE III. For Example III, the kernel g was replaced by a rectangular function of width 20. g is thus still smeared out, but since the Fourier coefficients of h do not decrease so rapidly, the problem is not so severely ill-posed.

Figure IIIA shows the mean square errors as functions of λ. They are minimized for different values of λ, but the discrepancy is not so great as in Example I.

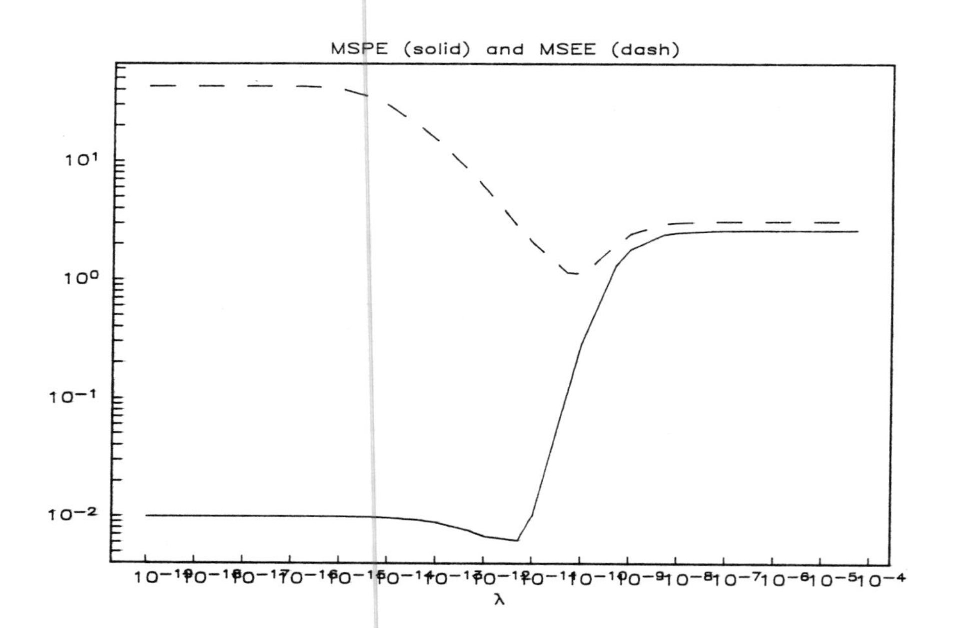

Figure IIIA.

Figure IIIB shows $\hat{f}$ and $\hat{g}$ when λ is chosen to be optimal for f. The shape of g can be discerned, although more smoothing is clearly called for.

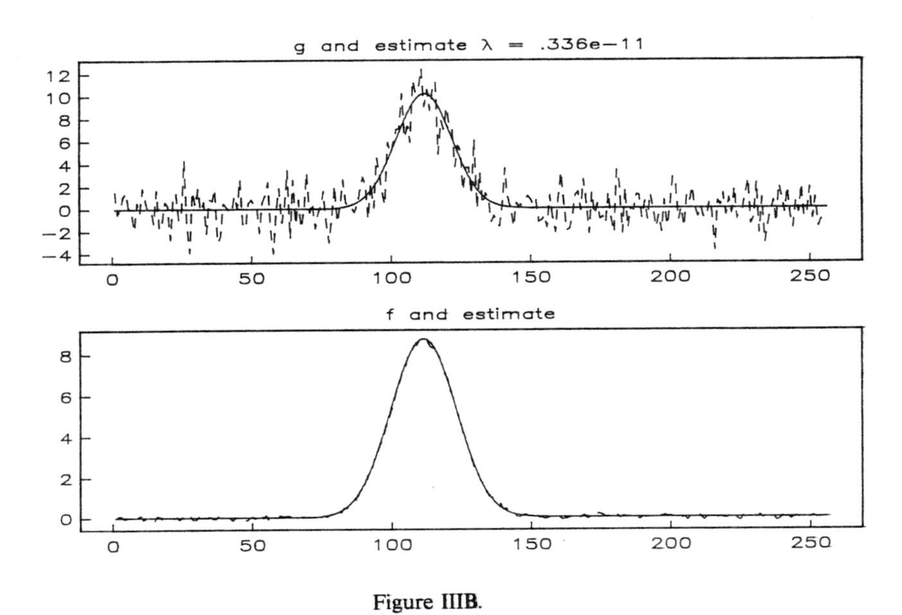

Figure IIIB.

Figure IIIC shows the estimates when λ is chosen to minimize *MSEE*. $\hat{g}$ is still quite rough, perhaps because of a Gibbs' effect, and f is again oversmoothed relative to the data.

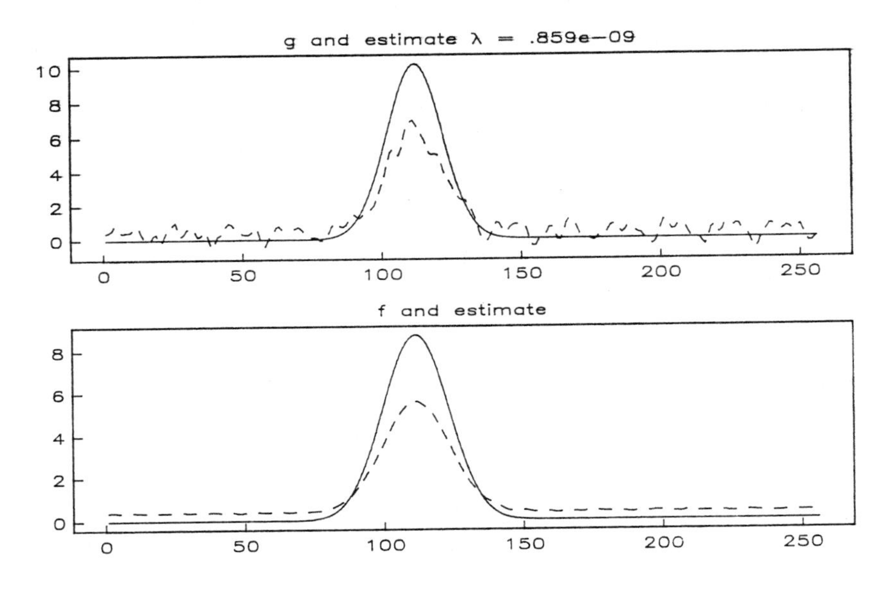

Figure IIIC.

EXAMPLE IV. In this example, the most difficult of the deconvolution problems, h was chosen to be a normal density with standard deviation 25 as in Example I, and g was chosen to be a doublet - the mixture of normal densities with standard deviations 10 and means 80 and 120.

Figure IVA shows that the structure of g is hopelessly lost when λ is chosen to minimize *MSPE*. In Figure IVB, where λ is chosen to minimize *MSEE*, we see that the estimate at best barely begins to resolve the doublet.

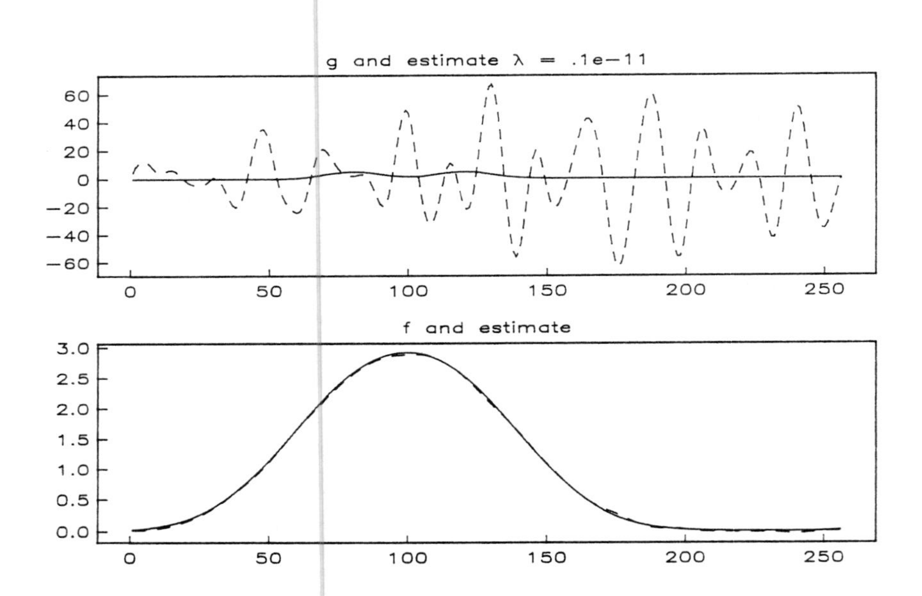

Figure IVA.

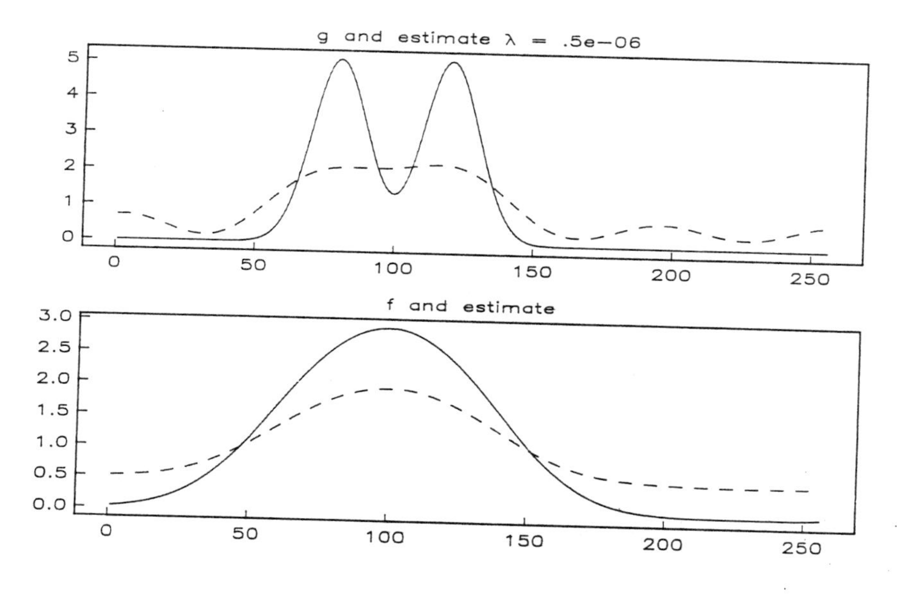

Figure IVB.

In all these examples we have seen the same basic phenomena. A reasonable smoothing of the observed data yields a deconvolution that is too erratic, and a smoothing parameter chosen to give a good deconvolution oversmooths the data.

5. CHOICE OF SMOOTHING PARAMETER FROM THE DATA

In nonparametric regression and density estimation there has been a great deal of theoretical and experimental work on the problem of choosing a smoothing parameter from the data. Typically, cross-validation or another procedure is used to estimate the risk or loss (usually squared error) as a function of the smoothing parameter, producing a curve from which a hopefully reasonable and maybe close to optimal value of the smoothing parameter is chosen. Analogues of these procedures exist for deconvolution problems as well, although their properties are even less well understood and there is not a lot of empirical evidence on performance to date.

I will first outline a scheme for choosing λ to minimize *MSPE*. Suppose that it is desired to choose λ to minimize the prediction loss function

$$P(\lambda) = \sum_{k=0}^{n-1} |f_k - \hat{f}_k|^2$$

$$= \sum_{k=0}^{n-1} |f_k|^2 - 2\mathrm{Re} \sum_{k=0}^{n-1} \hat{f}_k \bar{f}_k + \sum_{k=0}^{n-1} |\hat{f}_k|^2 .$$

The basic idea (see, for example, Rudemo (1982), Hall (1983), Stone (1984)) is simple: the first term does not depend on λ and the last term can be computed. The second term is estimated by finding an unbiased or nearly unbiased estimate of its expectation, and its expectation is clearly,

$$2\sum_{k=0}^{n-1} |f_k|^2 \frac{|h_k|^2}{|h_k|^2+\lambda k^4}.$$

Since

$$E\sum_{k=0}^{n-1} \left|\frac{y_{kn}}{n}\right|^2 \frac{|h_k|^2}{|h_k|^2+\lambda k^4}$$

$$= \sum_{k=0}^{n-1} |f_k|^2 \frac{|h_k|^2}{|h_k|^2+\lambda k^4} + \frac{\sigma^2}{n}\sum_{k=0}^{n-1} \frac{|h_k|^2}{|h_k|^2+\lambda k^4},$$

an unbiased estimate is

$$2\sum_{k=0}^{n-1} \left|\frac{y_{kn}}{n}\right|^2 \frac{|h_k|^2}{|h_k|^2+\lambda k^4} - 2\frac{\sigma^2}{n}\sum_{k=0}^{n-1} \frac{|h_k|^2}{|h_k|^2+\lambda k^4}.$$

Thus λ is chosen to minimize

$$P^*(\lambda) = -2\sum_{k=0}^{n-1} \left|\frac{y_{kn}}{n}\right|^2 \frac{|h_k|^2}{|h_k|^2+\lambda k^4} + 2\frac{\sigma^2}{n}\sum_{k=0}^{n-1} \frac{|h_k|^2}{|h_k|^2+\lambda k^4} + \sum_{k=0}^{n-1} |\hat{f}_k|^2.$$

In this analysis it has been assumed that σ^2 is known, which is typically not the case. However, σ^2 can be estimated from local fluctuations as in Rice (1983).

A similar process can be carried out for estimation error, as was pointed out to me by Paul Speckman. Suppose that the loss function for estimation is

$$E(\lambda) = \sum_{k=0}^{n-1} |g_k - \hat{g}_k|^2$$

$$= \sum_{k=0}^{n-1} |g_k|^2 - 2\mathrm{Re}\sum_{k=0}^{n-1} \hat{g}_k \bar{g}_k + \sum_{k=0}^{n-1} |\hat{g}_k|^2.$$

Again, the first term does not depend on λ, the third term can be computed, and the second term is estimated by an unbiased estimate of its expectation. Since

$$E(\bar{g}_k \hat{g}_k) = |g_k|^2 \frac{|h_k|^2}{|h_k|^2+\lambda k^4}$$

and

$$E\left\{\left|\frac{y_{kn}}{n}\right|^2 \frac{1}{|h_k|^2+\lambda k^4}\right\} = |g_k|^2 \frac{|h_k|^2}{|h_k|^2+\lambda k^4} + \frac{\frac{\sigma^2}{n}}{|h_k|^2+\lambda k^4}$$

instead of minimizing $E(\lambda)$, we minimize

$$E^*(\lambda) = -2\sum_{k=0}^{n-1} \left|\frac{y_{kn}}{n}\right|^2 \frac{1}{|h_k|^2+\lambda k^4} + 2\frac{\sigma^2}{n}\sum_{k=0}^{n-1} \frac{1}{|h_k|^2+\lambda k^4} + \sum_{k=0}^{n-1} |\hat{g}_k|^2.$$

Figures VA and VB show these criterion functions for a realization of the setup of Example I. The shapes are qualitatively reasonable, and in fact the minimizing λ over the grid on which the functions were computed are quite close to the theoretically optimal values of λ. This is not apparent from the figures because of the scaling - note that, because of the occurence of negative values, the y-axes are linear, not logarithmic as in Figure IA.

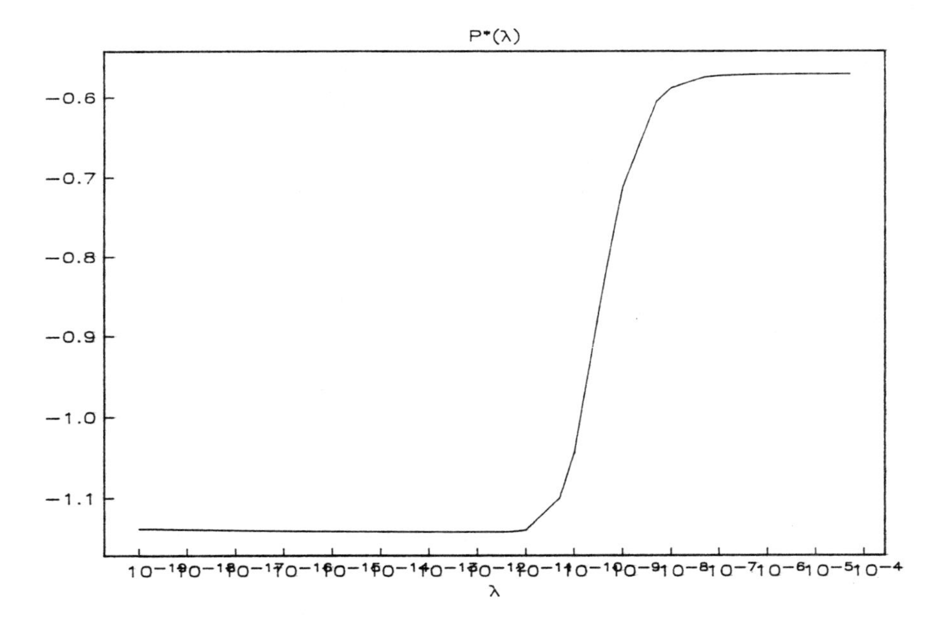

Figure VA.

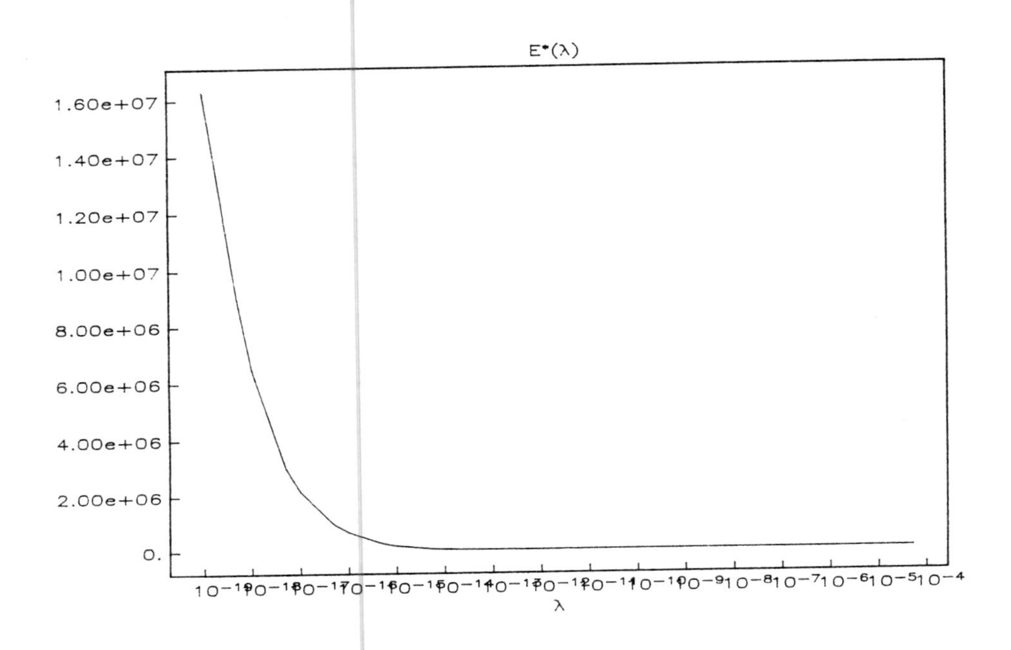

Figure VB.

6. CONCLUDING REMARKS

I have tried to point out, largely by example, that the choice of smoothing parameter should depend on whether one is interested in estimation or prediction and that a satisfactory choice for one aim may well be unsatisfactory for the other. A similar problem occurs in non-parametric differentiation (Rice, 1984); it is easy to see that if a smooth non-negative kernel is used to estimate a regression function and that if the bandwidth is chosen optimally for that purpose, then the estimate of the second derivative of the function is inconsistent.

Cross-validation, generalized cross-validation, and several variants have been employed quite successfully in non-parametric regression problems where there is little or no distinction between estimation and prediction. In inverse problems, of which the deconvolution problems I have considered here are a simple example, the emphasis of cross-validation on prediction makes it a less useful tool for estimation. The procedure outlined in the previous section and other ideas appear promising, but much empirical and theoretical work needs to be done.

REFERENCES

1. P. Hall, "Large sample optimality of least squares cross-validation in density estimation," Ann. Statist. **11** (1983), 1156-74.

2. J. Rice, "Bandwidth choice for nonparametric regression," Ann. Statist. **11** (1983), 1225-40.

3. J. Rice, "Bandwidth choice for nonparametric differentiation," J. Mult. Anal., (1984), to appear.

4. M. Rudemo, "Empirical choice of histograms and kernel density estimators," Scand. J. Statist. **9** (1982), 65-78.

5. C. Stone, "Window selection in density estimation," Ann. Statist. **12** (1984), 1285-97.

DEPARTMENT OF MATHEMATICS
UNIVERSITY OF CALIFORNIA, SAN DIEGO
LA JOLLA, CALIFORNIA 92093

Contemporary Mathematics
Volume 59, 1986

REGRESSION APPROXIMATION USING PROJECTIONS AND ISOTROPIC KERNELS

David L. Donoho
Iain M. Johnstone[1]

ABSTRACT. Projection pursuit regression (PPR) and kernel regression are methods for estimating a smooth function of several variables from noisy data obtained at scattered sites. When and by how much do projection methods reduce the "curse of dimensionality"? We focus on the two–dimensional problem, and study the L_2 approximation error ("bias") of the two proecdures with respect to Gaussian measure. A duality for polynomials extends to show that PPR-type approximations converge significantly faster than the minimax rate on radial functions and the Kernel-type approximations improve over the minimax rate on harmonic functions. The results extend to angular and harmonic 'smoothness classes' respectively. The two effects are complementary: they cannot occur together. The effects moderate as the underlying measure changes through a one parameter family to the uniform measure on a disc.

1. INTRODUCTION. Projection Pursuit Regression (Friedman and Stuetzle 1981, Friedman, 1985) is an addition to an array of multivariate smoothing methods that includes kernel and nearest neighbor regression, methods based on multivariate splines and orthogonal series and methods that produce step function fits such as recursive partitioning regression.

The idea guiding projection pursuit is that one can look for structure in multivariate data by searching through lower dimensional projections. A broad survey and history of such methods is given by Huber (1985). In the regression application, the idea of Friedman and Stuetzle was to try to reduce the (hard) multivariate smoothing problem to that of combining a number of appropriately chosen one dimensional smooths in which the responses $\{y_i\}$ were smoothed against a particular projections $\{\theta \cdot x_i\}$ of the independent variables $x_i \in \mathbb{R}^d$. A simpler, faster smoothing algorithm with smaller bandwidth can be used in the univariate smooth although some form of search over directions θ is needed.

In aiming to produce a fit to a regression function $f(x)$ having the form

---

1980 Mathematics Subject Classification Primary 62J02; Seconday 62H99, 41A10, 41A25, 42C10.

[1] Supported by NSF Grants DMS 8451750, 8024649.

$$\hat{f}_s(x) = \sum_1^s g_j(\theta_j \cdot x) \tag{1.1}$$

where $\theta_j \in S^{d-1}$ and $g_j : I\!R \to I\!R$ is a fitted non-linear 'ridge' function of a linear combination, the PP regression problem has some formal similarities to the reconstruction problem in X-ray computed tomography (e.g. Shepp and Kruskal, 1978, Davison and Grünbaum, 1981) and our theoretical development has been influenced by this connection. In practice projection pursuit regression deals with more variables, noisier data from a wider variety of sources and fewer directions - which are found by numerical search rather than being fixed in advance - than does X-ray CT, so in general far less precise results are obtained.

Nevertheless, early experiments with PPR reinforced the intuition that PPR's use of one-dimensional smooths might render it less vulnerable to Bellman's 'curse of dimensionality' that afflicts other multivariate regression methods. For example, if certain of the independent variables have no explanatory power, then the chosen projections can simply ignore them by assigning those variables zero coefficients.

This paper is an informal account of an approach we have taken to an approximation-theoretic question: which functions $f(\underline{x})$ can be well approximated by a (relatively parsimonious) fit of the projection pursuit type (1.1). We have had to make a number of limiting assumptions in order to make progress, the restriction to two independent variables and ignoring sampling variability being the two most serious. The assumptions and results are described more or less as presented at the Arcata meeting, detailed statements, proofs and discussions of most results can be found in (Donoho and Johnstone, 1986). No attempt is made here to address many of the other questions concerning PPR such as choice and performance of fitting algorithm, interpretation and inference for the fitted directions and functions.

In order to carry out the calculations, we work with a noise-free infinite data model with only two independent standard gaussian carriers. Thus, assume that $Y = f(\mathbf{X})$, where $\mathbf{X} = (X_1, X_2)$ has a standard Gaussian distribution in $I\!R^2$. (A one-parameter family of non-Gaussian carrier distributions is described in Section 5.)

Quality of approximation is measured by mean square error:

$$\| f - \hat{f} \|^2 = E[f(\mathbf{X}) - \hat{f}(\mathbf{X})]^2 = \int [f(\mathbf{x}) - \hat{f}(\mathbf{x})]^2 d\Phi_2(\mathbf{x}). \tag{1.2}$$

or, when appropriate, by relative error:

$$\| f - \hat{f} \| / \| f \| .$$

Remark: Consider a sampling model in which n i.i.d. observations are taken satisfying

$$Y_i = f(X_{i1}, X_{i2}) + \varepsilon_i$$

where $\{\mathbf{X}_i\}$ are distributed i.i.d. as Φ_2 and $\{\varepsilon_i\}$ are also i.i.d. and independent of $\{\mathbf{X}_i\}$. Then the mean integrated squared error suffered by an estimator $\hat{f}(\mathbf{x}) = \hat{f}(\mathbf{x}; \mathbf{X}_1, \ldots, \mathbf{X}_n, Y_1, \ldots, Y_n)$

decomposes into the sum of variance and squared bias terms. The squared bias term will have the form (1.2) if $\hat{f}$ there is replaced with $E\hat{f}(x; \mathbf{X}_1, \ldots, \mathbf{X}_n, Y_1, \ldots, Y_n)$. Thus the results of this paper concern bias properties of projection and kernel-based procedures.

Consider 'ridge-function' (or 'plane wave') approximations to f of the form

$$f_s(\mathbf{x}) = \sum_{i=1}^{s} g_i(\underset{\sim i}{\tilde{\theta}}^t \mathbf{x}) \tag{1.3}$$

where $g_i : \mathbb{R} \to \mathbb{R}$ is a univariate function of the linear combination $\underset{\sim}{\tilde{\theta}}^t \mathbf{x}$, and $\underset{\sim}{\tilde{\theta}}^t = (\cos\theta, \sin\theta), \theta \in [0, 2\pi)$.

We define an (s-term) projection pursuit (PP) approximation to f by

$$\hat{f}_{s,PP} = \arg \min_{\substack{\theta_i, g_i \in L^2(\Phi_1) \\ i=1,\ldots,s}} \| f - f_s \|^2 .$$

Thus, $\hat{f}_{s,PP}$ is obtained by minimizing over both functions $g_1, \ldots, g_s$ and directions $\theta_1 \ldots, \theta_s$. If the directions $\theta_1, \ldots, \theta_s$ are held fixed, and only the functions $g_1, \ldots, g_s$ are varied, we speak of a projection approximation (PA) to f:

$$\hat{f}_{s,PA} = \arg \min_{\substack{g_i \epsilon L^2(\Phi_1) \\ i=1,\ldots,s}} \| f - f_s \|^2 .$$

The ultimate goal is to study the properties of the PP approximation. The minimization over directions makes this a difficult non-linear problem, so our approach has been to study the performance of projection approximations based on fixed equally spaced directions. The latter problem is linear and analytically tractable and gives a lower bound on the quality of the PP approximation.

In computing projections, averages are taken over hyperplanes: lines in this two dimensional setting. Thus the neighborhoods have infinite extent in one direction and zero extent in the orthogonal direction. A complementary concept is that of averaging over circular neighborhoods, which leads us to consider kernel approximations (KA) to an $L^2(\Phi_2)$ function f given by

$$\hat{f}_\sigma = f * \varphi_\sigma .$$

Here φ_σ denotes a circular Gaussian density with covariance matrix $\sigma^2 I_2$ and $*$ denotes convolution. More generally, one could use an arbitrary circularly symmetric kernel $\psi(|x|)$, scaled by a bandwidth parameter σ.

Particular attention will be paid to harmonic functions $f(x, y)$, which satisfy the Laplace equation $\Delta f = \partial^2 f/\partial x^2 + \partial^2 f/\partial y^2 = 0$ and radial functions $f(x, y) = F(r^2)$, where $r^2 = x^2 + y^2$. Harmonic functions have the important mean-value property: if $B(x)$ denotes a ball in $\mathbb{R}^2$ centered at x, then

$$\int_{B(x)} f(y)dy = f(x).$$

Thus kernel approximations using isotropic kernels are exact at harmonic functions: in particular $f * \varphi_\sigma \equiv f$.

2. AN EXAMPLE. A comparison of kernel and equi-spaced projection approximation applied to simple eighth-degree radial and harmonic polynomials illustrates a phenomenon that motivates the later development.

The radial polynomial f_R was taken to be J_{44} in the system to be described below, $f_R(x,y) = r^8 - 32r^6 + 288r^4 - 768r^2 + 384$ and the harmonic polynomial $f_H(x,y)$ was J_{80}, namely $Re\ z^8$, where $z = x + iy \in C$.

Best $L^2(\Phi_2)$ approximations to f_R and f_H using s equally spaced directions $\theta_i = i\pi/s, i = 0, 1, \ldots, s-1$, were computed by standard least squares methods (described in detail in DJ). This was done for $s = 1, \ldots, 8$, at which point perfect approximation is possible since any degree m bivariate polynomial can be expressed as a sum of univariate m^{th} degree polynomials in m equally spaced directions. Relative errors were computed from formula (3.1) below and displayed in Table 1.

Relative approximation errors due to smoothing with a Gaussian kernel with various bandwidths σ were computed from formula (3.2) below for both f_R and f_H, and the results displayed in Table 1.

It is strikingly apparent that PA (projection approximation using equally spaced angles) works much better for the radial polynomial than for the harmonic. Indeed the ridge function fit has negligible explanatory power until all eight directions are included. Conversely KA is ideally matched to harmonic polynomials and so works much better for them than for the radial function.

Subsequent sections explore the following issues raised by this example:

- why is attention focused on radial and harmonic functions
- does the duality extend to wider classes of functions
- how strongly do the results depend on the Gaussian assumption for $\mathbf{X}$?

3. THE ROLE OF RADIAL AND HARMONIC FUNCTIONS. Consider fitting a single ridge function in (1.1). It is easily seen that the conditional expectation of f along (given) $\theta \cdot x$ yields the best fit:

$$\hat{f}_1(x) = P_\theta f(x) = E(f|\theta \cdot X = \theta \cdot x).$$

The benefit of assuming that $X \sim \Phi_2$ is that P_θ maps polynomials into polynomials of the same degree. Therefore orthogonally decompose

$$L^2(\Phi_2) = \mathcal{H}_0 \oplus \mathcal{H}_1 \oplus \ldots \oplus \mathcal{H}_m \oplus \ldots$$

where $\mathcal{H}_m$ is the space of bivariate Hermite polynomials of degree exactly m. Thus, if $H_k(x) = e^{x^2/2}(-\partial/\partial x)^k e^{-x^2/2}$ is the kth univariate Hermite polynomial, then $\mathcal{H}_m =$ span $\{H_j(x)H_{m-j}(y);\ 0 \le j \le m\}$.

For which function in $\mathcal{H}_m$ does projection pursuit perform least well? Define $PVE^*(f) = \sup_\theta \| P_\theta f \|^2 / \| f \|^2$. ("$PVE$" = "percent variance explained"). The questoin reduces to finding the minimum of $PVE^*(\cdot)$. Having found a minimum, one can search for a new minimum subject

relative error in PA approximation		
directions	harmonic	radial
1	0.996086	0.8523863
2	0.992157	0.6731456
3	0.994100	0.4313311
4	0.986013	0.1666666
5	0.991189	0.0
6	0.982607	0.0
7	0.942809	0.0
8	0.0	0.0

Table 1.1 **Relative error $\| f - \hat{f}_s \| / \| f \|$ for best ridge function approximation $\hat{f}_s$ using s equally spaced directions $\theta_i = i\pi/s$ $i = 1, \ldots, s$ to harmonic and radial polynomials of degree 8 (re $J_{8,0}$, $J_{4,4}$ respectively).**

relative error in KA approximation		
bandwidth	harmonic	radial
0.05	0	0.01000007
0.10	0	0.04000450
0.15	0	0.09005127
0.20	0	0.1602882
0.25	0	0.2511005
0.30	0	0.3632921
0.35	0	0.4983263
0.40	0	0.6586378
0.45	0	0.8480341
0.50	0	1.072221

Table 1.2 Relative error $\| f - \hat{f}_\sigma \| / \| f \|$ for kernel approximation $\hat{f}_\sigma = f * \varphi_\sigma$ using a Gaussian denisty with bandwidth σ for harmonic and radial polynomials of degree 8 (Re $J_{8,0}$, Re $J_{4,4}$ respectively).

to being orthogonal to the first. Proceeding thus as in a principal component analysis, one finds a sequence of two dimensional eigenspaces (except for the last, which is one-dimensional if m is odd). It is convenient to define complex valued polynomials $J_{0,m}, J_{1,m-1}, \ldots, J_{m/2,m/2}$ of degree m whose real and imaginary parts form bases for the successive eigenspaces. It is shown in DJ that we may take, on writing $x = r\cos\theta$ and $y = r\sin\theta$,

$$J_{k,m-k}(x,y) = c_{km} e^{i(m-2k)\theta} r^{m-2k} L_k^{m-2k}(r^2/2),$$

where $c_{km} = 2^k k!$ and $L_n^\alpha(x)$ is a Laguerre polynomial of degree $\alpha + 2n$ (Szegö, 1939, Ch. 5).

The mth degree polynomial that is worst for 1-term PP is $J_{0,m}(z) = z^m$, where $z = x + iy$. This displays the harmonic polynomials Rez^m and Imz^m as the basis for the worst eigenspace. As the other extreme, when m is even, the generator of the final one-dimensional subspace is $J_{m/2,m/2}$ which is a (real) radial function.

Aside from revealing the role of radial and harmonic polynomials, the $\{J_{k,l}\}$ basis is convenient for both projection and kernel approximation calculations. A simple example is

$$PVE^*(J_{k,m-k}) = \binom{m}{k} 2^{-m}.$$

An explicit formula can be given for the best s-term ridge function approximation using equally spaced directions $\theta_i = (i-1)\pi/s, i = 1, \ldots, s$ to the function $J_{k,m-k}$. Suppose that W_m is a Binomial $(m, 1/2)$ random variable, and that $PVE_s(f)$ denotes $\| f_s \|^2 / \| f \|^2$, the percentage variance 'explained' by the best s-term equi-spaced ridge function approximarion $P^{(s)} f$ to f. Then

$$\begin{aligned} (3.1) \qquad 1 - PVE_s(J_{k,m-k}) &= \frac{\| P^{(s)} J_{k,m-k} - J_{k,m-k} \|^2}{\| J_{k,m-k} \|^2} \\ &= Pr(W_m \neq k | W_m \equiv k[s]). \end{aligned}$$

(The initially surprising occurrence of conditioned random walk probabilities is a natural consequence of the Fourier analysis permitted by the use of equally spaced directions.)

The formula (3.1) is a tool for investigating whether parsimonious representations are possible with projection approximations. Now any bivariate polynomial of degree m has $0(m^2)$ coefficients. Furthermore, it can be represented as the sum of $m+1$ ridge polynomials each of degree m in $m+1$ (arbitrarily chosen) directions. The parsimony question becomes: is good approximation possible using m^{th} degree ridge polynomials is $s \ll m$ directions, as this involves only sm coefficients?

Figure 1 illustrates the answer for the $\{J_{k,m-k}\}$ basis provided by (3). The conditioning set $\{w : w \equiv k\ [s]\}$ is superimposed on a normal approximation to the density of W_m for large m. It is clear that until s increases to m, the amount of variance of a harmonic function explained remains (essentially) zero. However, for radial functions ($k = m/2$), as soon as $s = 0(\sqrt{m})$ quite good approximation is possible, requiring in the process only about $m^{3/2} (\ll m^2)$ coefficients. There is a continuous transition between the radial and harmonic extremes: this is shown in Figure 2.1: the contours are lines of constant relative error $\| P^{(4)} J_{k,l} - J_{k,l} \| / \| J_{k,l} \|$ as k and l vary ($s = 4$ ridge directions were used).

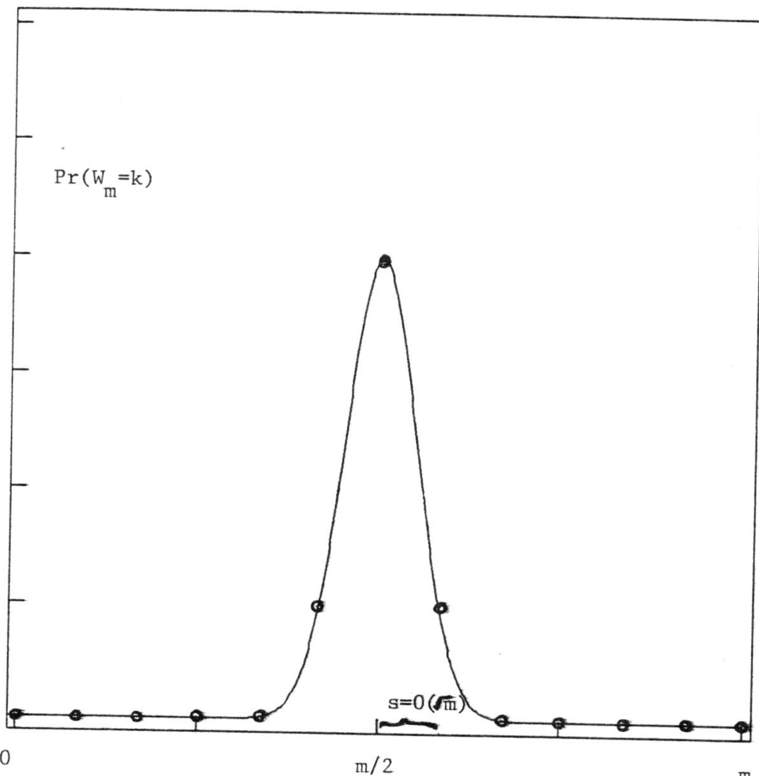

Figure 1 : $1 - PVE_s(J_{k,m-k}) = Pr(W_m \neq k | W_m \equiv k[s])$. The amount of variance explained by a fit using $s \equiv 0(\sqrt{n})$ directions is essentially negligible for a harmonic, but substantial for a harmonic function.

An equally explicit formula is available for the relative error in smoothing $J_{k,l}$ with a Gaussian kernel of bandwidth σ:

$$\text{(3.2)} \qquad \frac{\| J_{k,l} * \phi_\sigma - J_{k,l} \|^2}{\| J_{k,l} \|^2} = \sum_{j=1}^{k \wedge l} \binom{k}{j} \binom{l}{j} \sigma^{4j}$$
$$\sim kl\sigma^4 \quad \text{as } \sigma \to 0.$$

If attention is restricted to $J_{k,m-k}$ of total degree exactly m, then the relative error is maximised when $k = l = \frac{m}{2}$, that is for radial functions. Of course, the relative error vanishes for harmonic polynomials because of the mean value property.

A KA contour plot, Figure 2.2, showing lines of constant relative error in the (k, l) plane (for bandwidth $\sigma = .2$), contrasts sharply with the PA plot.

4. SMOOTHNESS CLASSES AND RATES OF CONVERGENCE. The aim in this section is to push the radial/harmonic dichotomy beyond the special case of polynomials and to seek broader qualitative descriptions of when PA and KA each achieve superior approximation.

Our formulation is guided by the following generalities. Let $\mathcal{C}$ be a class of (smooth) functions. An approximation method A such as PA, splines or polynomials for functions in $\mathcal{C}$ can be described

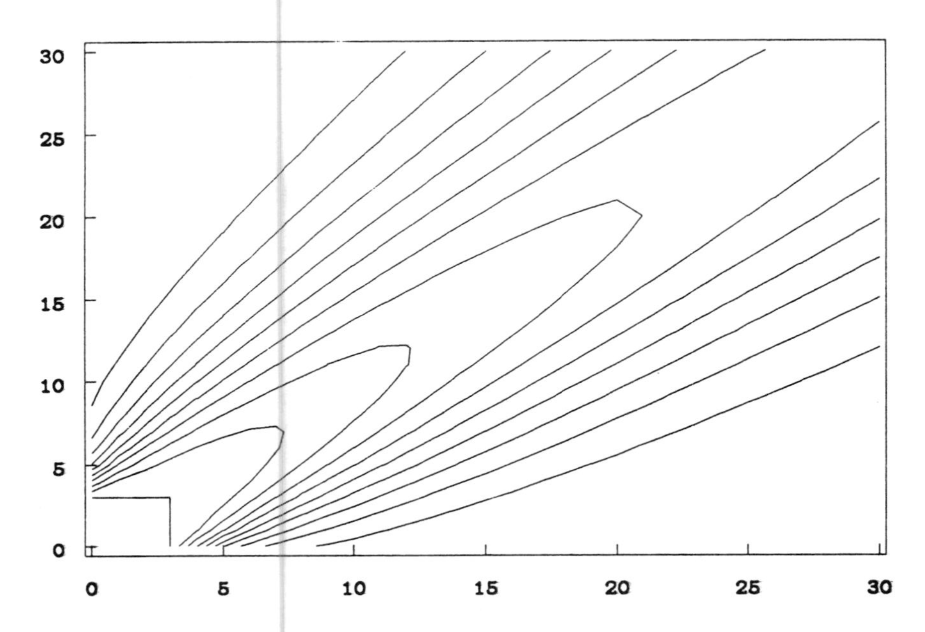

Figure 2.1 *PA* approximation error. Contour plot of relative approximation error $\| P^{(4)} J_{k,l} - J_{k,l} \| / \| J_{k,l} \|$ where $P^{(4)} J_{k,l}$ is the best approximation to $J_{k,l}$ using ridge functions in $s = 4$ equally spaced directions $\theta_i = \pi i/4$.

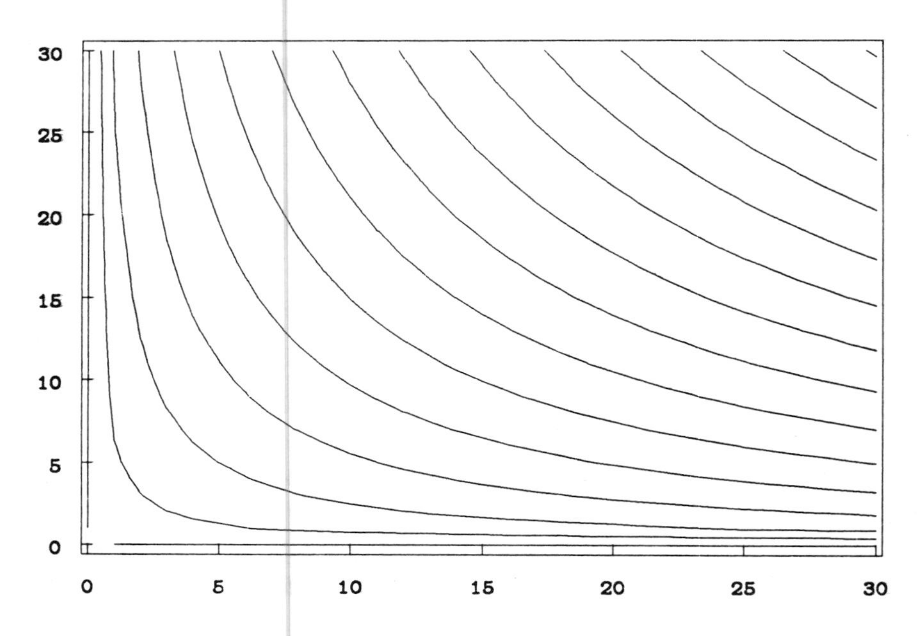

Figure 2.2 *KA* approximation error. Contour plot of relative approximation error $\| \hat{J}_{k,l} - J_{k,l} \| / \| J_{k,l} \|$ where $\hat{J}_{kl} = J_{kl} * \varphi_\sigma$ and $\sigma = .2$.

by a sequence of operators A_N of rank N, so that we think of $A_N f$ as using N coefficients to approximate f. A typical situation is that $\| A_N f - f \|^2 \sim N^{-r(f,A)}$, and that over all of $\mathcal{C}$

$$\sup_{f \in \mathcal{C}} \| A_N f - f \|^2 \sim N^{-r^*(\mathcal{C},A)}.$$

A minimax rate of convergence for the class $\mathcal{C}$ is the smallest number $r^*(\mathcal{C})$ which is at least as large as all rates $r^*(\mathcal{C}, A)$. For $L^2(\Phi_d)$-functions with an upper (norm) bound B on their p^{th} weak L^2 derivative, the minimax rate of convergence is $r^*(\mathcal{C}) = p/d$ (not the $2p/d$ that occurs for a uniform distribution on a compact set). Denoting this latter functions class by $\mathcal{F}_p^B$, our question becomes to describe subclasses of $\mathcal{F}_p^B$ on which KA or PA offers improvements over the minimax rate of convergence.

Remark: $\mathcal{F}_p$ will denote the linear space of $f \in L^2(\Phi_2)$ having p L^2 weak derivatives.

An illustrative result follows from Hoeffding's inequality applied to the (conditional) binomial probability in (3.1). Although $r^*(\mathcal{F}_p^B, PA) = p/2$, a faster rate of $r^d(\mathcal{F}_p^B \cap$ {radial functions}, $PA)$=1.5 is possible for the subclass of radial functions. (The rate includes a $\log N$ term not explicitly shown here: a careful statement is in DJ).

We now define classes of smooth functions intended to qualitatively capture the notions of "close to harmonic" or "close to radial". Since the degree of smoothness of a function can be measured by the rate of decay of its Fourier coefficients, consider the expansion of an $L^2(\Phi_2)$ function f in the form $\sum_{k,l} c_{kl} J_{kl}$. To describe decay conditions on the coefficients $\{c_{kl}\}$, introduce a probability measure (corresponding to f) on the positive integer lattice Z_2^+:

$$P_f(\{(k,l)\}) \propto |c_{kl}|^2 \| J_{kl} \|^2 .$$

where the constant of proportionality is chosen to produce a probability measure. Because differentiation acts very simply on the J_{kl} basis, it turns out that a necessary and sufficient condition for f to have p weak $L^2(\Phi_2)$ derivatives is (using the probability measure P_f defined above)

$$E_f(K+L)^p < \infty.$$

Radial functions are constant on circles centered at the origin, whereas harmonic functions achieve their maximum oscillation on the boundary of any disk. In particular the real and imaginary parts of $J_{0,n}$ oscillate m times between maximum and minimum going aroung a circle centered at 0.) This suggests that functions $f \in L^2(\Phi_2)$ be expressed in polar co-ordinates and the size of (weak) angular derivatives $\partial^q/\partial\theta^q$ be measured. Define $\mathcal{A}_{pq}$ as the linear subspace of $\mathcal{F}_p$ of functions for which $\partial^q f/\partial\theta^q$ lies in $L^2(\Phi_2)$. It is convenient that $\frac{\partial}{\partial\theta}(J_{kl}) = i(l-k)J_{kl}$, for this allows us to describe the class $\mathcal{A}_{pq}$ as those for which

$$E_f|K-L|^{2q} < \infty \quad \text{(as well as } E_f(K+L)^p < \infty).$$

To complement the classes of angularly smooth functions $\mathcal{A}_{pq}$, define $\mathcal{L}_{pr}$ as the linear subspace of $\mathcal{F}_p$ of functions for which the rth-iterated Laplacian $\Delta^r f \in L^2(\Phi)$. Since $\Delta J_{kl} = 4kl J_{k-1,l-1}$, this condition is equivalent to

$$E_f\{(KL)^r|K \geq r\} < \infty \quad (\text{as well as } E_f(K+L)^p < \infty).$$

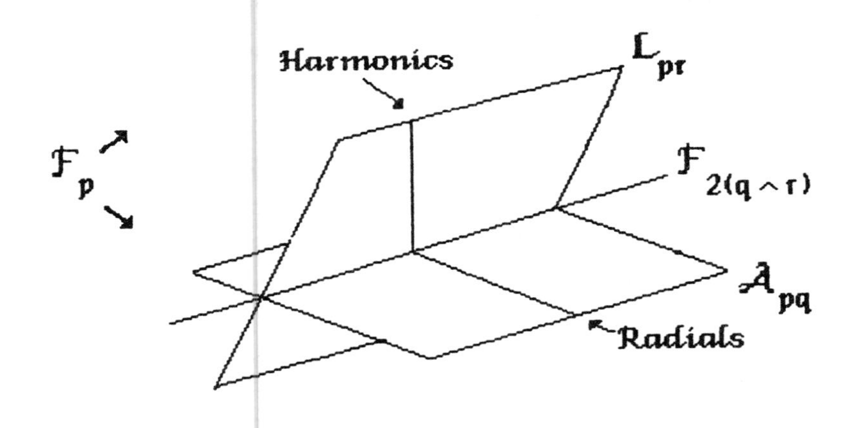

Figure 3. Inclusion relations amongst spaces of functions with "Cartesian" smoothness of order p $(\mathcal{F}_p)$, "angular" smoothness of order $q(\mathcal{A}_{pq})$ and Laplacian smoothness of order $r(\mathcal{L}_{pr})$.

Figure 3 shows the relationship between the spaces $\mathcal{F}_p, \mathcal{A}_{pq}, L_{pr}$. As is evident from the moment conditions, "angularly smooth" functions concentrate their spectral mass about the diagonal $k = l$ of radial basis functions, while the "Laplacian smooth" functions have spectral mass mostly distributed near the axes $k = 0$ or $l = 0$ corresponding to harmonic basis functions.

It is easy to check that if f has p L^2 derivatives then it automatically has $p/2$ angular and $p/2$ Laplacian derivatives. The moment conditions, together with the suggestive identity $(k + l)^2 = (k - l)^2 + 4kl$, indicate that higher amounts of angular and Laplacian smoothness may be complementary. Indeed it can be shown that if $f \in L^2(\Phi_2)$ has exactly p L^2 derivatives, then f cannot have both more than $p/2$ angular derivatives in L^2 and more than $p/2$ Laplacians in L^2.

In the following theorems $\mathcal{A}^B_{p,q}$ and $\mathcal{L}^B_{p,r}$ denote norm bounded subsets of $\mathcal{A}_{p,q}$ and $L_{p,r}$ respectively. Let $PA_{m,s}$ denote projection approximation using m^{th} degree polynomials in s equally spaced directions, and $f_{N,polys}$ denote the best bivariate polynomial approximation of degree at most $\sqrt{2N}$.

Theorem (Projection Approximation). Consider functions having p angular derivatives. There exist choices $m = m_N, s = s_N$ (with $ms = N$) for which

$$\sup_{\mathcal{A}^B_{p,p}} \| \hat{f}_{N,PA_{m,s}} - f \|^2 \leq c(\frac{N}{\log^2 N})^{-p/1.5} \tag{4.1}$$

whereas

$$\sup_{\mathcal{A}^B_{p,p}} \| \hat{f}_{N,polys} - f \|^2 \asymp N^{-p/2}.$$

For $\mathcal{A}^B_{p,q}$ with $p/2 \leq q < p$, a similar result holds with the right side of (4.1) replaced by $CN^{-(1+q/2p)}$.

Thus projection approximation is able to take advantage of enhanced angular smoothness to achieve as great an improvement in rate of approximation as is attained for radial functions. The next theorem gives a complementary result for kernel smoothing. Let $\psi(|x|)$ be an isotropic kernel on $\mathbb{R}^d$ (we have been taking $\alpha = 2$ above, but here this is not necessary) having compact support, $\int \psi dx = 1$ and $\int |x|^{2i} \psi dx = 0$ for $i = 1, \ldots, r-1$. Denote $\psi_\sigma(|x|) = \sigma^{-d}\psi(|x|/\sigma)$ and $f_\sigma = f * \psi_\sigma$. $\mathcal{L}^B_{p,r}$ denotes a norm bounded subset of $\mathcal{F}_p$ for which the L^2 norms of $\Delta^r f$ and $\Delta^r f(x \pm e_i)$ where $e_i = (0, \ldots, 0\, 1, 0 \ldots, 0)$ are also bounded by a fixed constant.

<u>Theorem</u> Consider functions having angular smoothness r. Then

$$\sup_{\mathcal{L}^B_{p,p}} \| f_\sigma - f \|^2 \leq c\sigma^{4p},$$

whereas

$$\sup_{\mathcal{F}^B_p} \| f_\sigma - f \|^2 \asymp \sigma^{2p}.$$

For $\mathcal{L}^B_{pr}$ with $p/2 \leq r \leq p$, a similar result holds with σ^{4p} replaced by σ^{4r}.

In summary, these results exhibit what might be fancifully thought of as a non-parametric analog of a phenomenon occurring in James-Stein estimation of a multivariate normal mean. In that setting, if there is prior information as to the location of the vector of unknown means, an estimator can be constructed that, while being minimax, achieves a substantial reduction in risk over the part of the parameter space near the prior mean. In the present non-parametric regression setting, KA and PA provide methods which while retaining a minimax rate over $\mathcal{F}_p$ can achieve substantial rate reductions over the subclasses $\mathcal{L}^B_{p,r}$ and $\mathcal{A}^B_{p,q}$ respectively (for $q, r \in (p/2, p]$).

5. NON-GAUSSIAN PHENOMENA FOR PROJECTION APPROXIMATION.

Suppose now that $X = (X_1, X_2)$ is uniformly distributed on the disk $\{x_1^2 + x_2^2 \leq 1\}$. The authors were initially surprised to discover that $PVE_s(J_{k,m-k})$ was virtually the same for all $k = 0, 1, \ldots, m$, so that the striking dichotomy between harmonic and radial functions disappears.

Fortunately, there is a one parameter family of measures on $\mathbb{R}^2$ connecting the uniform and Gaussian measures and having in addition the critical property that conditional expectation P_θ maps polynomials into polynomials of the same degree.

Define densities $f_X = f_{X,\lambda,R}$ by

$$f_X(x) = \frac{\lambda}{\pi \mathbb{R}^2}(1 - |x|^2/R^2)^{\lambda-1} \quad \text{for } |x| \leq R, \lambda \geq 1.$$

When λ is a positive integer, f_X is the density of the position of the closest to the origin of λ i.i.d. points chosen from $\{|x|^2 \leq I\!R\}$ according to a uniform distribution. Thus if $\lambda = 1, f_X$ is the uniform density, while if $R = \sqrt{2\lambda} \to \infty$, f_X converges to the standard Gaussian density on $I\!R^2$.

Much of the structure expressed by the J_{kl} basis in the Gaussian case has analogs for the λ-family and can be used to study how the properties of projection approximation change with decreasing λ. The use of the λ-family, as well as elements of the approach taken below, was suggested by articles of Davison (1981) and Davison and Grünbaum (1981) studying mathematical questions raised by X-ray computed tomography.

Let $X \sim f_{X,\lambda,R}$. Conditional expectation can be expressed as an operator P, $(Pf)(\theta, s) = E(f(X)|\theta \cdot X = s)$ mapping $L^2(I\!R^2, \mu_X)$ to $L^2([0,\pi] \times I\!R, \lambda \times \mu_{X_1})$, where λ denotes one dimensional Lebesgue measure and μ_X and μ_{X_1} the distributions of X and its marginal X_1 respectively. The singular value decompositon of PP^* (where P^* is the adjoint operator to P) produces eigenvalues λ_i and corresponding eigenfunctions g_i of PP^*. The functions $f_i = \frac{1}{\sqrt{\lambda_i}} P^* g_i$ form an orthonormal basis for the orthogonal complement of the kernel of P and are in fact precisely (multiples) of the J_{kl} basis functions introduced earlier from other considerations. Furthermore, there is the representation $Pf = \Sigma\sqrt{\lambda_i}(f, f_i)g_i$, from which one finds that f is hard to fit/reconstruct from Pf if f loads heavily into those basis functions f_i for which λ_i is small (the harmonic functions in the earlier treatment).

Davison and Grünbaum show that in the case of the λ-family, all eigenfunctions of PP^* have the form $e^{i\alpha\theta}\phi_m(s)$, where $\{\phi_m\}$ form an orthonormal basis for $L^2(I\!R^1, \theta^t X)$ (yielding Hermite polynomials in the Gaussian case, Chebychev Type II polynomials if $\lambda = 1$, and Gegenbauer polynomials in the general λ case). The corresponding eigenvalues are $\mu_{\alpha m}$, where

$$\mu_{m-2k,m} = \binom{m}{k} \frac{(\lambda)_k (\lambda)_{m-k}}{(2\lambda)_m} \text{ and } (\lambda)_k = \frac{\Gamma(\lambda + k)}{\Gamma(\lambda}.$$

We then define

$$J_{k,m-k} = \frac{c_m}{\mu_{m-2k,m}} P^*(Y_{m-2k}\phi_m) \quad (\text{where } Y_\alpha(\theta) = e^{i\alpha\theta})$$

where the particular value of c_m is unimportant but might be chosen for consistency with the Gaussian case as $R = \sqrt{2\lambda} \to \infty$. If $P_k^{\lambda,\alpha}$ denote Jacobi polynomials (Szegö, Ch. 4) it can be shown that

$$J_{k,m-k}(z) = c_{\lambda,k,m} e^{i(m-2k)\theta} r^{m-2k} P_k^{\lambda-1,m-2k}\left(\frac{2r^2}{R^2} - 1\right),$$

so that again the real and imaginary parts of $J_{0,m}$ are harmonic, while (for m even), $J_{m/2,m/2}$ is a radial polynomial of degree m.

Just as in the Gaussian case, there is a helpful probability interpretation of the singular values $\mu_{m-2k,m}$, at least when λ is an integer. Randomly order $2\lambda - 1$ white balls and m black balls and let W_m equal the number of black balls preceeding the middle (i.e. λ^{th} white ball). Then

$$\mu_{m-2k,m} = Pr(W_m = k)$$

and it is easily seen that this distribution has a mode at $m/2$ (unique if $\lambda > 1$), and tends to the binomial distribution as $\lambda \to \infty$ and the uniform on $\{0, 1, \ldots, m\}$ as $\lambda \to 1$.

As in the Gaussian case, projection pursuit with a single direction leads to

$$PVE^*(J_{k,m-k}) = \mu_{m-2k,k} = Pr(W_m = k).$$

More relevant is the analog of (3): if m^{th} degree polynomials in each of s equally spaced directions are used, then (3) is exactly reproduced (with the new definition of W_m):

$$1 - PVE_s(J_{k,m-k}) = P(W_m \neq k | W_m \equiv k\ [s]).$$

The same pictorial interpretations as for Figure 1 are now possible, as long as the new distribution of W_m is inserted. Thus one can ask how many directions s are now needed for good approximation of a bivariate radial polynomial of degree m. In the Gaussian case $s = 0(\sqrt{m \log m}) \ll m$ sufficed, but of course now the answer depends on λ. If $m = 0(\lambda^\alpha)$ for $\alpha \leq 1$ as $\lambda \to \infty$, then $s = 0(\sqrt{m \log m})$ continues to work, but if $m \asymp \lambda^\alpha$ for $\alpha > 1$ as $\lambda \to \infty$, then $s = 0(n^\rho)$ is needed, where $\rho = \rho(\alpha)$ is an as yet undetermined constant, but is known to lie in $(1/2, 1]$. One can interpret these results as asserting that the Gaussian dichotomy persists for the compactly supported distributions in the λ-family so long as the radial polynomials are relatively 'simple', i.e. have degree at most on the same order as the rate of decay of the tails of the X density.

Presumably these results can be extended to wider smoothness classes of functions under $L^2(\mu_X)$, as similar moment descriptions of smoothness classes and complementarity results are possible. These are summarized schematically below:

f has p weak $L^2(\mu_X)$ derivatives $\Leftrightarrow E_f[(K+L)\lambda + 2KL]^p < \infty$

f has q weak L^2 angular derivatives $\Leftrightarrow E_f(K-L)^{2q} < \infty$

f has r weak L^2 Laplacians $\Leftrightarrow E[[KL(K+\lambda)(L+\lambda)]^r | K \geq r] < \infty$, and

$$\{(K+L)\lambda + 2KL\}^2 = (K-L)^2\lambda^2 + 4KL(K+\lambda)(L+\lambda).$$

6. TENTATIVE CONCLUSIONS.

This section collects briefly some points from the earlier discussion that appear to deserve emphasis. In one-dimensional smoothing problems, all smoothing methods work by downweighting higher frequencies. For two or more independent variables, there are many qualitatively different ways for this to occur. For example, one can very roughly describe polynomial, projection and kernel approximation as working by downweighting coefficients in the $\{J_{kl}\}$ expansion that lie outside the regions in the (k, l) plane indicated below:

It is for this reason that the various methods will be better suited to functions whose spectral energy is concentrated in the shaded regions.

For approximation of functions by sums of ridge functions, the results for the Gaussian and lambda-family taken together might be tentatively paraphrased as follows. If oscillation of the target function f in the tails of the carrier distribution does not dominate the L^2 norm of the function, then projection approximation can improve over the minimax approximation rate.

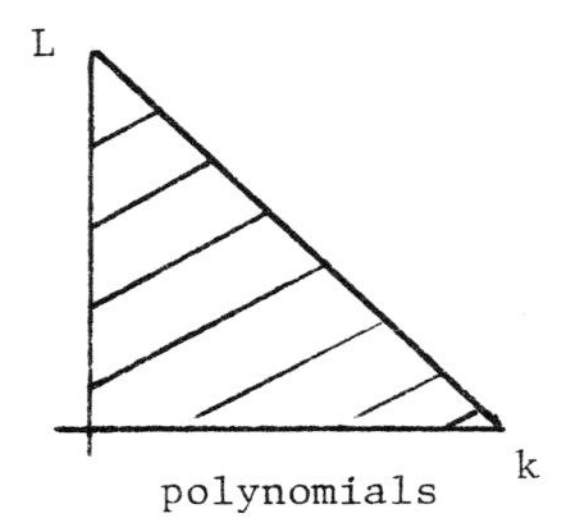

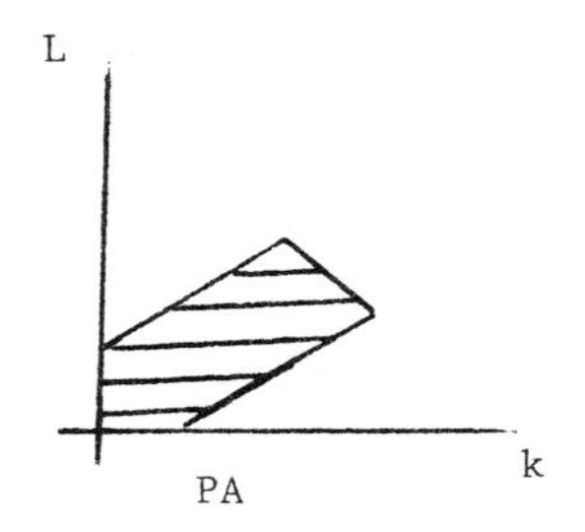

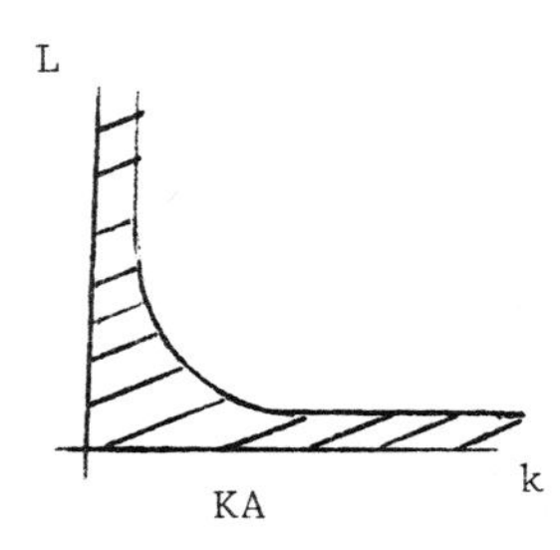

It might be interesting to identify function subclasses to which other regression procedures (such as the various multivariate spline methods) are particularly suited, thus obtaining (at least conceptually) information to aid in the choice of a smoothing procedure for a particular multivariate problem.

Among the many issues left unaddressed here, and which we hope to pursue in future, are

- behavior in higher dimensions, the setting for which PP methods are chiefly envisaged
- behavior for more general distributions on the independent variables (especially those having heavier tails than the Gaussian)
- when are rates faster than 1.5 possible in the two dimensional problem
- how will these results go over into a sampling setting, in which both functions (and perhaps) directions are estimated from a finite amount of data
- to what extent are the results tied to the use of an L^2 error criterion. For example, kernel smooths can produce qualitatively superior fits than ridge approximations even when the latter have smaller L^2 errors.

BIBLIOGRAPHY

1. Davison, M.E. and Grünbaum, F.A. (1981). Tomographic Reconstruction with Arbitrary Directions. *Comm. Pure Appl. Math*, **34**, 77-120.

2. Davison, M.E. (1981). A singular value decomposition for the Radon Transform in n-dimensional Euclidean space. *Numer. Funct. Anal. and Optimiz.*, **3**(3), 321-340.

3. Donoho, D. and Johnstone, I.M. (1986). Projection Based Smoothing and a Duality with Kernel Methods. To appear *Ann. Statist.*.

4. Friedman, J.H. (1985). Classification and multiple regression through projection pursuit. L.C.S. technical report # 12, Department of Statistics, Stanford University.

5. Friedman, J.H. and Stuetzle, W. (1981). Projection pursuit regression. *J. Amer. Statist. Assoc.*, **76**, 817-823.

6. Huber, P.J. (1985). Projection pursuit (with discussion). To appear *Ann. Statist.*.

7. Shepp, L.A. and Kruskal, J.B. (1978). "Computerized Tomography: The new Medical X-ray Technology." *Am. Math. Monthly*, **85**, 420-439.

8. Szegö, G. (1939). Orthogonal Polynomials, Vol. 23 of *American Mathematical Society Colloquium Publications*, American Mathematical Society, Providence, R.I..

Department of Statistics
University of California
Berkeley, CA 94720

Department of Statistics
Stanford University
Stanford, CA 94305

Contemporary Mathematics
Volume 59, 1986

WILL THE ART OF SMOOTHING EVER BECOME A SCIENCE?

J. S. Marron[1]

ABSTRACT. The general problem of smoothing parameter selection is considered in the specific setting of kernel density estimation. Recent results on smoothing parameter selection are put into a framework analogous to the classical development of the theory of nonparametric function estimation. The serious consequences of these results are discussed.

1. INTRODUCTION. Essentially all nonparametric function estimation techniques can be thought of as smoothing operations. Their effective performance is critically dependent on the choice of a smoothing parameter. If not enough smoothing is done, the resulting estimate exhibits features which are not part of the curve being estimated, but instead are artifacts of the particular set of data at hand, ie. the variance is quite large. If too much smoothing is done, important features of the true curve could be eliminated by "smoothing them away", ie. the bias is quite large..

Kernel density estimation, which is perhaps the purest and simplest kind of smoothing, is considered here. The goal is to recover a density function, $f(x)$, from a sample, $X_1,\ldots,X_n$, which comes from f. The estimator is given by

$$\hat{f}(x) = \sum_{i=1}^{n} \frac{1}{nh} K\left(\frac{x-X_i}{h}\right),$$

where K is called the kernel and is often taken to be a symmetric probability density, and where h is the smoothing parameter which is often called the bandwidth.

It is well known that the lessons learned in smoothing by the study of density estimation give a great deal of insight into what is happening in more complicated, but probably more important settings, such as spectral density and regression estimation. All the attention given to smoothing splines at this

1980 Mathematics Subject Classification (1985 Revision). 62G20.
[1] Supported by NSF Grant DMS-8400602.

workshop demonstrates quite well that there are some very attractive, but again quite complicated, alternatives to kernel estimators. However, in a first attempt at understanding some of the deeper issues involved with smoothing methods, the kernel density estimator seems appropriate because it is so much simpler from a technical point of view.

Section 2 provides a context for the automatic smoothing parameter selection results which are described in section 3. Sections 4 and 5 contain some extensions amd discussion. Section 6 gives conclusions.

2. THE CLASSICAL APPROACH. Until roughly 1982, most research on nonparametric function estimation, fell into one or more of three types of results: consistency, rates of convergence, and bounds on the rates of convergence.

Consistency results simply state that as the sample size grows, the estimate may be expected to get close to the curve being estimated in some sense. More precisely, if f is the curve to be estimated, if $\hat{f}$ is the estimator, and if $\|\cdot\|$ denotes some sort of error criterion such as such as an L^p norm or maybe mean square error at a point, then

$$(2.1) \qquad \|\hat{f} - f\| \to 0,$$

in some mode of convergence. There are far too many results of this type to cite them all, but perhaps the most elegant are those of Silverman(1978) and Devroye and Penrod(1984).

It is not surprising that most any reasonable estimator will be consistent. So given several consistent estimators one should look for some means of comparison. One possibility for this is to study the rate of convergence in (2.1). Such results are typically of the type

$$(2.2) \qquad \|\hat{f} - f\| \sim n^{-p},$$

in some sense, for some power p which often is determined by the smoothness assumptions, eg. number of derivatives, made on f and on how $\hat{f}$ is constructed. The most famous results of this type are those of Rosenblatt(1956), Parzen(1962), Watson and Leadbetter(1963), and Rosenblatt(1971). It is interesting to note that these were essentially the first papers on smoothing methods and were published before a large number of papers which only deal with the weaker notion of consistency. The power p is nearly always less than 1/2, which is not surpising because $n^{-1/2}$ is the rate one could expect to get by assuming a parametric form for f and estimating the parameters. This difference between the parametric and nonparametric will be seen to appear in a very interesting and surprising way in section 4. An important special case is kernel density estimation with K a symmetric probability density and with f twice continuously differentiable. Then, if $h \sim n^{-1/5}$, p is 2/5.

The bound type of result was motivated by the observation that, for a given amount of smoothness of f, p was the same for quite a few reasonable estimators. See Wahba(1975) for a nice demonstration of this. Hence, it seemed reasonable to suspect that there might be some limit on p in terms of the smoothness of f. The proof of the existence of such a limit, in the minimax sense, is provided by the bound type of theorem. The first of these is Farrell(1972). There are a good number of others, maybe the most elegant being Stone(1980) and Stone(1982). A way of formulating such results will now be given. Let $\hat{f}$ now be any measurable function of the data, so as to take into account both estimators which have already been proposed and also those that may some day be invented. To hope to be able to say the rate of convergence in (2.2) is not too fast, a mechanism for ruling out trivialities like $\hat{f} = f$ is required. Hence, a minimax approach is used. Let $\mathcal{F}$ denote a suitable class of functions, then p is a bound on the rate of convergence in the sense that

$$(2.3) \qquad \sup_{f \in \mathcal{F}} P_f[\|\hat{f} - f\| > cn^{-p}] \rightarrow 1,$$

for c small enough in some sense.

3. AUTOMATIC BANDWIDTH SELECTION. While results of the above type give a good deal of insight into what is happening in estimation by smoothing methods, they unfortunately are not very useful for practical selection of the smoothing parameter. Hence there has recently been a good deal of work on data based methods for choosing the smoothing parameter. Among the most promising methos for bandwidth selection in kernel density estimation is least squares cross-validation, which was introduced by Rudemo(1982) and Bowman(1984), and may be motivated as follows.

A very compelling choice of the bandwidth is the minimizer of the Integrated Squared Error,

$$ISE(h) = \int [\hat{f} - f]^2.$$

By expanding the square, ISE can be written as a sum of three terms, one of which is independent of h, and the other two of which can be well estimated by

$$CV(h) = \int \hat{f}^2 + n^{-1} \sum_{j=1}^{n} \hat{f}_j(X_j),$$

where $\hat{f}_j$ denotes the kernel estimator constructed by leaving X_j out of the sample. Hence, it seems reasonable to use $\hat{h}$, the minimizer of CV(h), as the bandwidth in estimating f by $\hat{f}$.

The rest of this section describes recent results which investigate the properties of $\hat{h}$ in a way which can be thought of as being quite analogous to the way in which the results summarized in section 2 study the properties of $\hat{f}$. First a decision must be made as to what shall be called the optimal h. The fact that this choice is not so easy will be demonstrated shortly. The notion of optimal that appears most often in the literature is h_0, the minimizer of the Mean Integrated Squared Error,

$$\mathrm{MISE}(h) = E[\mathrm{ISE}(h)].$$

However, it is proposed to take the optimal h to be $\hat{h}_0$, the minimizer of $\mathrm{ISE}(h)$, because $\hat{h}_0$ is the bandwidth which makes $\hat{f}$ as close to f as possible for the data set at hand, instead of over all possible data sets.

In papers on smoothing parameter selection, results which are analogs of the consistency results of (2.1) have been given the name "asymptotic optimality". They are typically formulated as

$$\frac{\mathrm{ISE}(\hat{h})}{\mathrm{ISE}(\hat{h}_0)} \to 1, \tag{3.1}$$

or sometimes as

$$\hat{h}/\hat{h}_0 \to 1, \tag{3.2}$$

in some mode of convergence. The first result of this type was by Hall(1983), and there are a number of other such results in the literature, perhaps most notably by Stone(1984), who probes deeply into what is driving asymptotic optimality by using conditions even weaker than those required for the consistency of $\hat{f}$ to f.

While asymptotic optimality is a comforting fact, it only says that there exist sample sizes that are large enough that $\hat{h}$ will behave like $\hat{h}_0$. An important question is then: how large a sample size is required before one may be confident that the asymptotics are really describing the situation? Following the spirit of section 2, one way to approach this is to study rates of convergence in (3.1) and (3.2). Hall and Marron(1985a) have shown that, in what is essentially the case of K a symmetric probability density with f twice continuously differentible,

$$n^{1/10}\left(\frac{\hat{h}-\hat{h}_0}{\hat{h}_0}\right) \to N(0,\sigma^2), \tag{3.3}$$

(3.4) $$n(ISE(\hat{h})-ISE(\hat{h}_0)) \rightarrow C\cdot\chi^2_1,$$

in distribution, for constants C and σ^2. While these results are stated in terms of asymptotic distributions, note that a consequence of them is that they give a rate for the convergence in (3.1) and (3.2). A shocking feature is that the rate of convergence of the relative difference in (3.3) is so slow. Observe that to make $n^{-1/10}$ equal to 0.1, a ridiculously large sample of size 10 billion is required. Hence careful interpretation of asymptotic optimality is required.

At first glance, a possible interpretation of these results is that least squares cross-validation is not doing as well as it should. However, there is a sense in which it is doing surprisingly well. In particular, Hall and Marron(1985a) have also shown that

(3.5) $$n^{1/10}\left(\frac{\hat{h}_0-h_0}{h_0}\right) \rightarrow N(0,\sigma_0^2),$$

(3.6) $$n(ISE(\hat{h}_0)-ISE(h_0)) \rightarrow C_0\cdot\chi^2_1,$$

in distribution, for some other constants σ_0^2 and C_0. One consequence of this result is that least squares cross-validation is doing very well in that the order of its error is the same as the order of the relative difference between $\hat{h}_0$ and h_0, although it can be shown that $\sigma^2 > \sigma_0^2$. Another consequence is that the question of what to take as the optimal h is more serious than previously believed. In particular, for finite samples, there can be a surprisingly large relative difference between $\hat{h}_0$ and h_0.

A further consequence of this result is that one should be very careful about using "plug in" methods of bandwidth selection. These were first proposed by Woodroofe(1972), and involve plugging estimates into an asymptotic representation of h_0. The chief drawback of these methods is that the quantities to be estimated are typically even harder to estimate than f itself. However, even if this difficulty were not present, and one knew those quantities exactly, the resulting bandwidth could still not be expected to be substantially closer to the optimum than the cross-validated $\hat{h}$.

From the fact that the excuciatingly slow $n^{-1/10}$ rate of convergence of $\hat{h}$ to $\hat{h}_0$ is the same as the rate of convergence of $\hat{h}_0$ to h_0, one might again take a cue from the spirit of the ideas of section 2, and be lead to suspect that this rate may be the best that can be hoped for. The fact that this is indeed the case, in a minimax sense very similar to that of (2.3), has been demonstrated by Hall and Marron(1985b). In particular, it has been shown, under suitable assumptions, for $\hat{h}$ any function of the data,

$$(3.7)\qquad \sup_{f\in\mathcal{F}} P_f[|\frac{\hat{h}-\hat{h}_0}{\hat{h}_0}| > cn^{-1/10}] \rightarrow 1,$$

$$(3.8)\qquad \sup_{f\in\mathcal{F}} P_f[(ISE(\hat{h}) - ISE(\hat{h}_0) > cn^{-1}] \rightarrow 1,$$

for c small enough in some sense. The consequence of this result is that, while least-squares cross-validation can be expected to give a bandwidth which is rather far from the optimum, there will never be a substantially better method of bandwidth selection.

4. EXTENSIONS. Observe that the rate of convergence results of section 2 were stated for a general power p, while in section 3 only a particular rate was considered. This was done to simplify the presentation. In fact all the results of section 3 can be generalized in two directions.

The first direction of generalization is where K is allowed to take on negative values. It is well known that, if K is chosen so that its first k-1 moments vanish, and if f has k continuous derivatives, then the rate of convergence in (2.2) is $p = k/(2k+1)$ (when $\|\cdot\|$ denotes a proper norm and not a squared error criterion). In particular, the bigger k gets, the faster the rate of convergence, and in the limit, the rate is the same as the parametric rate of $p = 1/2$.

Both the rate and the bound results of section 3 can be extended to cover this case. When this is done in (3.3), (3.5), and (3.7), the rate 1/10 becomes 1/(4k+2). In (3.4), (3.6), and (3.8), the ISE rate stays the same. This result is really rather paradoxical. It says that in situations where $\hat{f}$ can be expected to be closer to f, ie. k large, $\hat{h}$ will be farther away from $\hat{h}_0$. Furthermore, as the rate of convergence of $\hat{f}$ to f approaches the parametric rate of $p = 1/2$, $\hat{h}$ will be converging to $\hat{h}_0$ arbitrarily slowly.

During the course of this workshop, Dennis Cox said that he believes he has actually observed this last effect. He was doing some simulations involving kernel density estimation with the function sinx/x as kernel. This function is closely related to working in the case $k = \infty$ above. Dennis reported that cross-validation did a terrible job of bandwidth selection in that setting.

Also during the workshop, a very insightful remark was made by Paul Speckman which provides a way of thinking about the above paradox. Paul said he has the feeling that cross-validation is taking advantage of the difference between the parametric and the nonpararmetric rates of convergences. This fits in perfectly with the above results because, for k large, $(1/2)-k/(2k+1)$ is small. Note that this is still a long way from an explanation of why this should happen, but it is an interesting interpretation.

The other direction in which the results of section 3 may be extended is to the multivariate case. In particular, if the X_i are d-dimensional random vectors, then the rate in (2.2) becomes $p = 2/(4+d)$. Note that here, as d gets bigger it becomes harder to estimate f, and the rate can be arbitrarily slow when d is sufficiently large.

When the rate and bound results, (3.3) through (3.8), are extended to this case, 1/10 becomes $d/2(d+4)$ and once again the ISE rate stays the same. Hence, as d increases, $\hat{h}$ will tend to be closer to h. Note that this fits in with Speckman's interpretation perfectly, since f is harder to estimate, there is more room for effective selection of h.

A possible means of resolving this paradox, of h being easier to select when f is harder to estimate, comes from a remark made by Warren Sarle. In doing simulations in the case of rather large d, Warren observed that the sensitivity of the estimator increases rapidly with d. Hence, for large d, the increased demands on the accuracy of the selection of h may outweigh the fact that there will be less noise in $\hat{h}$. This idea may work the other way for large k, namely there may be less demands on accuracy, so more noise may be acceptable, although there does not as yet seem to be any evidence one way or the other.

Hans-Georg Müller has suggested a very nice way of understanding this analytically. He points out (see (16) of Gasser, Müller, and Mammitzsch(1985)) that, for general d and k,

$$\frac{d^2}{dh^2}\text{MISE}(h)\Big|_{h=h_0} \sim n^{-(2k-2)/(2k+d)}.$$

Note that for d=1 and k large, this shows that there is a relatively small amount of curvature in the curve MISE(h) at its minimum, so the minimum is harder to find in the presence of noise. On the other hand, for k=2 and d large, this effect goes the other way. Observe that this idea both casts insight into what causes the above paradox, and also appears to quantify the effect observed by Sarle.

The reason that the rate of convergence in (3.4) does not change in either of these two cases is not yet understood. Since that rate is n^{-1} and the limiting distribution is chi square with one degree of freedom, it appears that what is going on is some sort of square of an ordinary central limit theorem, but a close inspection of the proofs has failed to turn up what is driving that central limit theorem.

5. EFFECT ON ESTIMATION. A question that was raised by Dennis Cox during the workshop was: what do these results imply about how close $\hat{f}$ is to f? In particular what can be said about $\hat{f}(x,\hat{h})$ (using obvious notation).

To see the answer to this, write

$$\hat{f}(x,\hat{h}) - f(x) = [\hat{f}(x,\hat{h}) - \hat{f}(x,h_0)] + [\hat{f}(x,h_0) - f(x)].$$

The behavior of the second term on the right is very well understood, and under the usual assumptions, tends to zero at the rate $n^{-2/5}$. By a Taylor expansion argument which is similar to, but simpler than, those in Hall and Marron(1985), the first term can be shown to be of the order $n^{-1/2}$. Hence, at least technically speaking, the first term is of lower order, and one may think of $\hat{f}(x,\hat{h})$ as being essentially the same as $\hat{f}(x,h_0)$. But it is important to keep in mind that the relative difference is again the terribly slow rate of $n^{-1/10}$, so it may take massive sample sizes before this is realistically true in practice.

6. CONCLUSION. It seems the most important conclusion to be drawn from these results is that while smoothing parameters which have been chosen by cross-validation are subject to a lot of noise, it is not possible to do much better. In particular, it looks like the common data analytic practice, of doing smoothing by looking at pictures for several different values of the smoothing parameter and taking the one that looks most pleasing, will not be entirely replaced by automatic smoothing methods. Also it looks like, while a scientist can gain a great deal of insight by using smoothing methods, he can never use them to convince a skeptic that the conclusions he draws are correct. Hence the answer to the question "will the art of smoothing ever become a science?" appears to be "no".

One might think that a consequence of these results is that they will have the effect of putting out of business people who study smoothing from a theoretical standpoint. The fact that this is probably not true has been brought out by a remark made by Charles Stone. He said that one way to interpret these results is that they indicate that one should be careful to guard against thinking too much in terms of error criteria. Some recent simulations seem to bear this out. In particular, these simulations appear to indicate that what cross-validation is really doing is something like finding the smallest bandwidth which gives a curve containing only features which should be present for that amount of smoothing. If this turns out to be true, then cross-validation could be a very useful tool for finding insights into particular data sets.

BIBLIOGRAPHY

Bowman, A., "An alternative method of cross-validation for the smoothing of density estimates", Biometrika, 65 (1985), 521-528.

Devroye, L., "A note on the L_1 consistency of variable kernel estimates", Ann. Statist., 13 (1985), 1041-1049.

Devroye, L. and Penrod, C.S., "The consistency of automatic kernel density estimates", Ann. Statist., 12 (1984), 1231-1249.

Farrell, R. H., "On the best obtainable rates of convergence in estimation of a density function at a point", Ann. Math. Statist., 43 (1972), 170-180.

Gasser, T., Müller, H-G. and Mammitzsch, V., "Kernels for nonparametric curve estimation", J. Royal Statist. Soc., Ser. B, 47 (1985), 238-252.

Hall, P., "Large sample optimality of least square cross-validation in density estimation", Ann. Statist., 11 (1983), 1156-1174.

Hall, P. and Marron, J. S., "Extent to which least-squares cross-validation minimises integrated square error in nonparametric density estimation", Center for Stochastic Processes Technical Report No. 94. (1985a), (submitted to Z. Wahrsch. Geb.)

Hall, P. and Marron, J. S., "The amount of noise inherent in bandwidth selection for a kernel density estimator", Center for Stochastic Processes Technical Report No. 100. (1985b), (submitted to Ann. Statist.)

Parzen, E., "On estimation of a probability density function and mode", Ann. Math. Statist., 33 (1962), 1065-1076.

Rosenblatt, M., "Remarks on some non-parametric estimates of a density function", Ann. Math. Statist., 27 (1956), 832-837.

Rosenblatt, M., "Curve estimates", Ann. Math. Statist., 42 (1971), 1815-1842.

Rudemo, M., "Empirical choice of histograms and kernel density estimators", Scand. J. Statist., 9 (1982), 65-78.

Silverman, B. W., "Weak and strong uniform consistency of the kernel estimate of a density and its derivatives", Ann. Statist., 6, 177-184. (addendum 8, 1175-1176)

Stone, C. J., "Optimal convergence rates for nonparametric estimators", Ann. Statist., 8 (1980),, 1348-1360.

Stone, C. J., "Op[illegible] etric regression", Ann. Stati[illegible]

Stone, C. J., "An [illegible] for kernel density estimates", [illegible]

Wahba, G., "Optim[illegible] ernel, and orthogonal series meth[illegible] (1975), 15-29.

Watson, G. S. and [illegible] probability density I", Ann. Mat[illegible]

Woodroofe, M., "O[illegible], 41 (1970), 1665-1671.

DEPARTMENT O[illegible]
UNIVERSITY O[illegible]
CHAPEL HILL, [illegible]

ABCDEFGHIJ --89876